WATER-WISE GARDENING

WATER-WISE GARDENING

Steve Solomon

GOOD BOOKS

Water-Wise Gardening
Steve Solomon

Good Books Publishing
goodbookspub.com

ISBN: 978-1-955289-10-8

Contents

Foreword   i

Introduction to the Good Books Edition   iii

Introduction   vii

Gardening With Rainfall

1   Gardening With Rain   3

2   Water-Wise Science   9

3   Minimizing Irrigation   19

4   Starting Seeds   47

5   How to Grow It: A-Z   55

6   My Own Garden Plan   123

Gardeners Textbook of Sprinkler Irrigation

7   General Principles   135

8   Systematic Watering   139

9   Designing Irrigation Systems   145

10 Uniformity of Application   151

11 Drip   159

12 Plant Spacing Possibilities   161

More Reading   167

Foreword

I first became acquainted with Steve Solomon's work back when I was gardening in Tennessee. My wife and I were broke, concerned about the financial crash and trying to make ends meet while doing everything possible to secure our family food supply. Steve's book *Gardening When it Counts* looked like a good read so I picked up a copy. As I read, I soon discovered that it wasn't just a good read—it was a work that would transform how I gardened.

At that point I had already been gardening for years, but some of the practical knowledge Steve taught filled in gaps in my practice I didn't even know were gaps. Before reading that book, I didn't know how to properly sharpen and adjust a garden hoe for my height. Or how combining fertilization with irrigation could improve crop yields in times of drought. Or that you could make a "dust mulch" to protect the ground from losing water. My wife read the book as well and Steve Solomon became our gardening mentor through a tough stretch. Some years later when he released *The Intelligent Gardener* with co-author Erica Reinheimer, I snatched it up and read it from cover to cover and discovered a whole new world of knowledge about the connection between plant nutrition and human health.

After I became a garden writer myself, I finally got to meet Steve through his Soil and Health community. We immediately struck up a friendship that has endured despite us being on opposite sides of the globe. As the author of ten gardening books and two gardening thrillers, I have appreciated the chance to tap into Steve's insight as I write.

In 2021, I left my previous publisher to start my own publishing house—Good Books. Knowing that Steve's excellent book *Water-Wise Gardening*, or *Gardening Without Irrigation*, had gone out of print, I asked if he would be interested in letting me republish it. To my delight, he not only agreed to let me get it back in print, he decided to re-write and expand the work into a volume that covers not only dry farming but the proper use of irrigation in the garden.

If you were raised on gardening books extolling the wonders of dense plantings in tiny beds, this book will open your eyes to an entirely new realm of knowledge. Our ancestors knew how to grow food without pressure-treated lumber, massive piles of mulch, fancy containers or garden sprinklers. I think most of us have never considered trying to garden on just rainfall—yet it's often possible. Sometimes it even gives you better-tasting and more nutrient-dense food.

In *Water-Wise Gardening*, you'll learn how to simply and effectively grow food using ancient time-tested techniques that consume less water and less fertilizer and may even take less effort. This second edition is almost a new book, and it's a valuable resource for gardening in good times and in bad, helping you steward your liquid assets while still getting the best-looking and best-tasting produce you can imagine. I've learned more with every reading and I believe you'll find it a valuable addition to your self-sufficiency library.

David The Good
Atmore, Alabama
2021

Introduction to the Good Books Edition

I wrote *Water-Wise Vegetables* in 1992–3. It was only distributed in a unique climatic region called Cascadia. The book reported on my own discoveries about growing vegetables with little or no irrigation in western Oregon, where there is no useful rainfall from June through September. Achieving that level of food independence interested few Cascadians of that era so *Water-Wise* soon went out of print. That did not matter much in the great scheme of things because reducing the need for irrigation became part of the warp and woof of everything I wrote thereafter.

Water-Wise Vegetables also encouraged Amy Garrett's Dry Farming Collaborative. Thanks to Amy's efforts there now are a goodly number of Cascadians doing distributed trials involving unirrigated food production. They collectively know more about the topic than I do. Isn't that good!

David Goodman (a.k.a. David The Good) is a food gardening writer who self-publishes and makes a decent living doing so. We have become friends. Earlier this year he launched his own publishing house, Good Books. David obtained a used copy of *Water-Wise Gardening*. He asked: would it be okay with me if he republished it? Nah, I said. Not as it was. I know a lot more now. So 28 years after publication of the first version I have completely rewritten the book and in the process, expanded it to include as much of North America as I can.

You should know about a few things that occurred in my life after I published *Water-Wise*. In 1994 my wife Isabelle and I moved from Elkton, Oregon to Nelson, British Columbia. Two years after that

Isabelle died. A scant year later I started hanging out in Australia and a year after that, 1998, I settled in Tasmania and have been here ever since.

Tasmania is about as far from the world as you can get while still enjoying a pleasant climate and a truly civil society. Country Tasmanians these days are a lot like rural Oregonians probably were in the 1950s. There's only half a million of us living on a temperate South Pacific island that is half the size of western Oregon. The climate is a lot like western Oregon. Most Taswegians are strongly connected to family. It is not easy to get wealthy here; many of our young people seek a wider set of options in the big smokes of Melbourne or Sydney.

Soon after settling here I discovered that Tasmanians faced food gardening difficulties much like Oregonians did when I first arrived there in 1978. Like Cascadia relates to the rest of the United States, Tasmania is the tail of a big dog (mainland Australia) that doesn't consider us important. No other part of Australia has a climate like ours. Mainland seed companies and mainland garden writers do not speak our language. So after I lived here a few years I wrote a book very much like my *Growing Vegetables West of the Cascades.* Not surprisingly it is called *Growing Vegetables South of Australia,* the title slyly suggesting that Tasmania should be an independent country.

I self-publish GVSOA. It is a labor of love. I print it in batches of 400 copies. The inside pages slowly grind out in my home office using the smallest office copy center FujiXerox sells. The covers and binding are done by a local printer. I sell around 600 copies a year and have done so since 2002. Nearly every time I print a batch I make minor improvements in the text because I keep learning how to garden better.

I always go the opposite way the crowd is moving. That is my nature. I am approaching age 79. My wife Annie is 81. One year ago we lived on a quarter acre suburban block. We also owned an adjoining vacant quarter acre block; it was our veggie garden. We grew more than half of all the food we ate year-round, which is why we are still flexible and capable of putting in a hard morning's work. We had an unlimited amount of municipal water. I used some of the water-wise techniques you will read about in this book but we mostly depended on sprinkler irrigation. Nine months ago we moved house—but not into a retirement village or care home as you might expect. Nah! We bought 11 acres of gently sloping grazing land in a microclimate where, as the locals put it, winter nights are three blankets colder than where we had been living.

The geology under our new property provides no possibility of a well. All our water—household and irrigation—comes off a roof and is stored in rainwater capture tanks. Like Oregon, Tasmania gets almost all its rainfall in winter, very little if any of it in summer. During a normal winter about 65,000 gallons flow through our water tanks. They hold about 26,000 gallons when the rains taper off in mid-spring. This is all the water we have to support a 2,500 sq. ft. food garden through summer and supply household needs as well.

Now you'll understand why I will use the past tense when referring to what I *did* in Oregon. And when I say what I do now, you'll know I mean in Tasmania.

This book divides into two parts. The first section is about gardening with little or no irrigation. A second short section is for gardeners with plenty of water. It explains how to irrigate scientifically and effectively.

In this book I will reference other worthy books. All but a few are classics few people know about any longer. Most of them have long been out of print. To make this valuable information readily available I created a website called The Soil And Health Library, soilandhealth.org There is no charge for downloading books. Any book title I refer to in *Water-Wise Vegetables* that is available for free download through the The Soil And Health Library will have an asterisk mark (*).

It is my prayer that this book helps North Americans withstand the challenges they are facing now and the far greater challenges they are almost certain to face in the next few years.

Steve Solomon
August, 2021

Introduction

Drouth is said to be the arch enemy of the dry-farmer, but few agree upon its meaning. For the purpose of this volume, drouth may be defined as a condition under which crops fail to mature because of an insufficient supply of water. Providence has generally been charged with causing drouths, but under the above definition, man is usually the cause. Occasionally, relatively dry years occur, but they are seldom dry enough to cause crop failures if proper methods of farming have been practised.

There are four chief causes of drouth: (1) Improper or careless preparation of the soil; (2) failure to store the natural precipitation in the soil; (3) failure to apply proper cultural methods for keeping the moisture in the soil until needed by plants, and (4) sowing too much seed for the available soil-moisture.

—John A. Widtsoe, *Dry Farming*

Irrigation hugely improves the garden's performance. Small salad radishes and celery quickly lose eating quality if they do not have moist soil at all times. Moisture-stressed onion seedlings hang on but do not grow. Irrigation can increase the yield from all vegetable crops. Irrigation always makes the outcome more certain. But many country dwellers must carefully ration a limited water supply; many urban and suburban dwellers may wish to grow food on a vacant block without convenient water or in their own back yard when their use of municipal water is restricted. These folks could grow much more food with knowledge of what I call dry gardening.

How I Got Interested in Dry Gardening

Tomato planting out time was approaching. My age was 36; the year was 1978. My wife and I had just settled on a five acre Oregon Coast Range homestead near Lorane, which is at the extreme limit of reasonable commuting distance for those who work in Eugene. We'd come there by way of Los Angeles, where I had spent the previous seven years earning and saving (barely) enough to homestead free-and-clear.

We moved on to the property early in May. In that season the Lorane Valley is covered with green grass; the days are punctuated by liquid sunshine and rainbows. New to a territory reputed for having rainy weather, I thought the grass would stay green all summer. I was wrong.

I intended to produce as much of my own food as possible, raise some kind of as-yet-to-be-determined (legal) crop to pay the bills, live simply so as to need little cash. Because I had horticultural ambitions, my offer to purchase had been subject to first drilling a well delivering at least 10 gallons per minute—at my expense. The fifth attempt (and what would be the final try) brought in 15 gpm. But I was drilling early in April and did not yet appreciate Cascadia's annual rainfall pattern. I bought the property and returned to Los Angeles to sell up and pack. A month later we moved into a large tent, spread crushed rock on the driveway, put in a septic tank and power pole and soon hauled in a new single-wide mobile home—not the rustic dream home we'd wanted but this was instant free-and-clear shelter. We had enough left in the bank to bare-bones subsist for two years if nothing major went wrong.

Then I set about establishing a 5,000 square foot food garden. I was not very smart about building soil fertility in those days so I spread an inch-thick layer of sawdusty horse manure, a few bags of lime and cottonseed meal, hired a neighbor to rototill the space, erected a deer fence and installed an excessively permanent irrigation system using 10 all brass agricultural impulse sprinklers that accepted the lowest output nozzles available. Each sprinkler threw only 1 gallon per minute. They uniformly spread slightly more than ⅛ inch of water per hour over the entire garden. I expected our 15 gallon per minute well would comfortably supply all 10 sprinklers at once and allow reasonable household use at the same time. All ten sprinklers were turned on and off with one valve located just outside the garden gate. I used galvanized pipes for supply lines and risers because I knew I was going to be there for the rest of my life. Wasn't I a clever young man! Actually, every time in this lifetime I thought that something would be so for the rest of my life I've turned out to be (usually painfully) wrong.

Then summer arrived in early June and those lush, green Lorane Valley pastures quickly browned off. It hadn't rained a drop for two weeks. I didn't yet realize that a normal Oregon summer brings warm to hot sunny days and zero rain from June through September. My garden needed irrigating for the first time. So I opened the valve and stood outside the garden gate admiring ten little impact sprinklers bouncing out crisscrossing streams, *psit, psit, psit!* But after a few minutes the pump shut down. Not yet knowing much about Oregon Coast Range wells and how they typically declined in summer, I spent a few hours repeatedly priming and restarting the pump before I realized that the pump kept shutting off because my tested and proven 15 gallon per minute well was delivering less than 10 gallons per minute.

In response, I ran fewer sprinklers at one time. I was profoundly worried—was I drawing from an underground stream that diminished in summer like most above-ground streams but would not disappear—or would it dry up altogether? Had we purchased a property that would not even feed us, much less grow something for sale? By late July the well could sustain only three sprinklers. Fortunately, that was its minimum output. I could still water the entire garden, but in stages. *Pfew!* With our food supply secured, I looked for a way to earn a homestead-based income. After going down several dead ends I started a mail order vegetable seed company. Then that pitiful well was also called upon to sustain Territorial Seed Company's half acre trials ground. Initially I watered the trials with only one 2½ gallon per-minute impact sprinkler mounted in a bucket filled with concrete, fed by a hose. It spread water at such a low rate over such a large area that I could only operate it when the sun was not shining. I'd start up the sprinkler about 8 pm, wake by alarm clock at 2 am, stumble barefoot down the cold dewy path, move the sprinkler to its next regular position, wipe off my wet feet, go back to bed and shut the system down at 8:00 am so the household could have morning showers. This went on four to five nights a week. No wonder that when the seed business could afford drip irrigation I started using it. With drip I could water the trials in the daytime and water the household garden at night.

I arranged the trials ground quite differently from how I grew our family food garden. The food garden produced a lot of food from a relatively small space. Not knowing better in those days I used the fashionable intensive method, with plants very close together on raised beds. These crowded plants had to be watered every few days. The trials ground was not intended to grow half the family's food supply; it produced information. So the plants were positioned far enough apart in the row that each one had all the space it could use. That way it would be obvious if all the plants from one variety were uniformly productive or highly variable. The spacing also ensured that weaker plants wouldn't be suppressed by strong ones close by. The trial rows were 100 feet long and four feet apart center to center. This spacing let me conveniently walk between rows while making observations and taking notes.

A few years later the well pump failed in the middle of summer. It took more than one week to replace it. After a few unwatered days the intensive kitchen garden plants began to complain, yet the widely spaced trials ground showed no moisture stress even after a week without watering. I believe I was being Shown how tenuous my food production system actually was. I became curious. How did the Oregon Territory's early settlers grow vegetables in a climate where most summers lacked useful rainfall during the entire frost-free growing season?

A century ago most rural homes depended upon hand- or wind-pumped wells. Sometimes there was a cistern just outside the kitchen window with a small hand pump next to the kitchen sink. American garden books of that era prescribed arranging vegetable crops single file in long rows, row centers three or four feet apart with narrow foot paths between rows—like my trials ground. Water was precious then.

Even dirty dish water became irrigation. Irrigation was doled out with a watering can. This worked in the Eastern United States and Canada because it rains frequently enough during summer. It worked in the states immediately west of the Mississippi River—Minnesota, Iowa, Missouri and Arkansas. It also worked in East Texas.

It worked because drought does not necessarily start after two weeks without rain. John Widtsoe, whose wise words begin this chapter, defines drought as beginning when the crop is severely damaged. The onset of drought has less to do with how much time elapses between rains than with how much water the soil naturally makes available to plants and upon how wisely the grower manages that moisture reserve.

Being in the vegetable garden seed business I heard about local gardening lore. There was an old guy growing unirrigated carrots on deep alluvial sandy loam in the hottest microclimate of Southwestern Oregon; this part of western Oregon also gets the least annual rainfall and has almost no hope of summer rain ever. He sowed carrot seed in spring when the soil was naturally moist enough to germinate it. He soon thinned the row to one foot apart; the single rows were four feet apart. He gave his carrots no irrigation whatsoever but they grew to enormous sizes, remained quite edible all summer, and the weight of good quality food harvested from the area involved was not as low as you might think.

I found further hints in a book by Gary Nabahan called *The Desert Smells Like Rain*, which describes Native Americans in Arizona growing unirrigated gardens on deep soil that retains moisture from a brief period of seasonal rains. They used extraordinarily wide interplant spacing. I read about how native South Americans in the Andean highlands grow food crops in a cool, low-humidity environment that gets only 12 inches of annual rain. And I discovered John Widstoe.

Gardening With Less Irrigation

In the spring of 1987 we moved to a 16-acre homestead with excellent soil and an abundant well that did not decline in summer. There I unintentionally grew my first summertime vegetable without any irrigation. I had just sold Territorial Seed Company and "retired" at age 44. Instead of reselling seeds that I had purchased from major seed producers I set out to grow my own seeds. One environmental factor that worsens germination and lowers seed vigor is damp conditions during the last weeks before harvest. So, in early March 1988, I moved six winter-surviving savoy cabbage plants far beyond the irrigated soil of my raised-bed vegetable garden. I transplanted them 4 feet apart because blooming brassicas make huge sprays of flower stalks. I did not plan to water these plants at all because cabbage seed forms during May and dries down during June as the soil naturally dries out.

That is just what happened. Except that one plant did something unusual. Instead of dying after setting a massive load of seed, this plant also threw a vegetative bud that grew a new cabbage among the seed stalks. With increasing excitement I watched this head grow steadily larger through the hottest and driest summer I had ever experienced. Realizing I was being gifted with a Revelation, I gave the plant absolutely no water although I did hoe out the weeds around it after I removed the seed stalks. I harvested the unexpected lesson in September. The cabbage weighed 6 pounds and was sweet and tender.

I had witnessed a demonstration of how elbow room might be the key to gardening with little or no irrigation. So I researched dry farming and soil/water physics. In the spring of 1989, I rototilled a test patch holding four widely separated, unirrigated experimental rows with five feet between row centers and a few feet of bare soil to either side of the patch. I tested an assortment of vegetable species spaced far apart in the row. To make the game more challenging I used absolutely no water, not even for sprinkling the seeds to get them germinating, which meant I had to sow everything in spring before the topsoil dried out.

I tried kale, savoy cabbage, Purple Sprouting broccoli, carrots, beets, parsnips, parsley, endive, dry beans, potatoes, French sorrel, and a few sweetcorn plants. I included one compact very early ripening bush (determinate) and one sprawling (indeterminate) tomato plant. I confess that these two tomato seedlings were given 1 cup of water when they were transplanted. Many of these vegetables grew surprisingly well. I harvested unwatered tomatoes from mid-July through September; kale, parsley, and root crops fed us through autumn and winter. The Purple Sprouting broccoli bloomed abundantly the next March.

In terms of quality, the harvest from all these vegetables (but one) was acceptable. The root vegetables were far larger but only a little bit tougher and quite a bit sweeter than usual. The potatoes yielded less than I'd been used to and had thicker than usual skin, but also had a better flavor and kept especially well through the winter. The rutabagas were a failure; they grew to the size of soccer balls but had been hollowed out by cabbage root maggots. You gardeners in the Eastern United States and Canada may have never heard of the root maggot—and you can count that as a blessing.

The following year I grew two gardens; each one occupied one-sixth acre. One of them, my "insurance garden," was thoroughly irrigated, guaranteeing we would have plenty to eat. An experimental plot of equal size was entirely unirrigated. There I tested species that I hoped could grow through a rainless summer.

By July, growth on some species had slowed to a crawl and they looked stunted. Wondering if a hidden cause of what appeared to be lack of moisture might actually be nutrient deficiencies, I tried spraying liquid fertilizer directly on these gnarly leaves, a practice called foliar feeding. It helped greatly because, I reasoned at the time, most fertility is located in the topsoil, and when the topsoil gets dry the plants draw on subsoil moisture. Plant nutrients I had put into the topsoil had become unobtainable. That being so, I reasoned that some of these species might do even better if they had just a little fertilized water. So I improvised a primitive drip system and metered out a few gallons to some of the plants in late July and a few more gallons around mid-August. Farmers call this method "fertigation." Unfertigated winter squash vines were small and scraggly and yielded about 15 pounds of food. Fertigated vines grew much larger, looked healthier and yielded 50 pounds. Thirty-five additional pounds of squash in exchange for 20 gallons of water and a bit of soluble nutrition seems a pretty good return on investment.

The next year I integrated all this experience into just one garden. Water-loving species like lettuce and celery were grown through the summer on a large, thoroughly irrigated raised bed. I hand watered this bed with hose and nozzle. The rest of the garden was given either no irrigation at all or else minimal carefully measured fertigations. Some unfertigated crops were foliar fed.

Everything worked in 1991! So, the next year, 1992, I set up a sprinkler system to water one central raised bed that was 100 feet long and four feet wide and used the overspray to support species that grew better with some moisture supplementation; I fertigated some plants located entirely beyond the sprinklers' throw while keeping a large section of the garden entirely unwatered. At the end of that summer I wrote the first version of this book. What follows is not mere theory, nor is it a summary of what I read about. These techniques are tested and workable.

Part I

Gardening With Rainfall

Chapter 1

Gardening With Rain

Before the mid-1970s, American garden books said to arrange most vegetables in long single rows with footpaths between each row. Between-row spacing for most vegetables usually ranged from three to four feet. The distance between rows had as much to do with the amount of anticipated summertime rainfall as it did with the type of vegetable. Sprawling crops like winter squash, cucumber and most tomato varieties were allowed even more space between rows. Small-sized vegetables like carrot and beet might be assigned a pair of parallel rows with a foot between them, the center of each pair of rows three or four feet to the next row with a footpath between them. In the eastern half of the United States and Canada all but a few kinds of vegetables can successfully forage for moisture when grown that way. In 1979 when I went into the vegetable garden seed business the average backyard food garden was 1,000 square feet. I think the average house lot prior to the mid-1970s was about 15,000 square feet, ⅜ acre. In small towns it was ½ acre.

Summertime rainfall usually comes often enough and in large enough quantity to be sufficient for vegetable gardening in any part of the United States where the native vegetation had been forest. The main use for irrigation was to help germinate seeds; it was done using hose and nozzle or a watering can. Once seedlings were up and growing, rainfall took care of it. When there was an unusually dry summer the vegetable garden might be watered a few times, usually with lawn sprinklers. Garden books of that era paid little or no attention to irrigation.

In the mid 1970s a new method called "intensive" was introduced by a West Coast guru named John Jeavons. His book *How To Grow More Vegetables Than You Ever Thought Possible On Less Land Than You Ever Imagined* has been through innumerable revisions and is still in print. Jeavons's basic idea is that by using wide super-fertile raised beds and decreasing interplant spacing as much as possible, yield from a given space could be greatly increased. My own experience with this system says this assertion is as much hyperbole as it is truth.

Jeavons's method was strongly promoted and soon was broadly accepted. By the mid-1980s every vegetable gardening book writer (except myself), every magazine article and every extension service publication intended for gardeners was regurgitating Jeavons's methods. I think the underlying conditions favoring Jeavons's narrative were population pressure, inflation and with it a steady erosion of real prosperity that was forcing new homes to be built on ever-smaller lots. I think it highly likely that The Powers That Be wanted the ever-harder-struggling average person to believe that a postage stamp garden could be as productive as a much larger one.

One thing about intensive food gardening is absolutely certain: the method *requires* frequent irrigation because densely planted beds can be sucked dry after a few hot sunny summer days. This might be okay as long as there is plenty of irrigation water available. It also makes the garden as needy of care as a pet can be when the family goes on holiday.

Rainfall During the Growing Season

Few places in the United States receive as much rainfall as the soil is losing during the hottest months of summer. Cascadia is an extreme example of that. Soil moisture loss in Cascadia averages 1½ inches per week during summer; rainfall during that period is next to nothing. This amount of moisture loss is roughly correct where a crop has grown large enough to shade the ground with leaves, so in Cascadia when generous irrigation is possible it makes sense to spread 1½ inches of water each week and even more during really hot weather. On soils with a fair bit of clay in them this much water could be added once a week. On light soils that do not hold as much moisture, a part of the weekly moisture loss should be added every two or three days.

Summertime Rainfall West of the Cascades (in inches)

Location	Apr.	May	June	July	Aug.	Sept.	Oct.
Eureka, CA	3.0	2.1	0.7	0.1	0.3	0.7	3.2
Medford, OR	1.0	1.4	0.98	0.3	0.3	0.6	2.1
Eugene, OR	2.3	2.1	1.3	0.3	0.6	1.3	4.0
Portland, 0R	2.2	2.1	1.6	0.5	0.8	1.6	3.6
Astoria, OR	4.6	2.7	2.5	1.0	1.5	2.8	6.8
Olympia,WA	3.1	1.9	1.6	0.7	1.2	2.1	5.3
Seattle, WA	2.4	1.7	1.6	0.8	1.0	2.1	4.0
Bellingham, WA	2.3	1.8	1.9	1.0	1.1	2.0	3.7
Vancouver, BC	3.3	2.8	2.5	1.2	1.7	3.6	5.8
Victoria, BC	1.2	1.0	0.9	0.4	0.6	1.5	2.8

How about the Eastern United States? It is convenient to divide the USA into two main areas by drawing a north/south line running through Dallas, Texas, which is the 98th meridian. West of that line the land grows grass which greens up in spring and browns off in summer. There are few trees, mostly adjacent to flowing water. The reason grass dominates and trees are scarce to non-existent is that severe drought happens often enough to kill trees. East of the 98th meridian the native vegetation was forest.

Summertime Rainfall Eastern USA (in inches)

Location	Apr.	May	June	July	Aug.	Sept.	Oct.
Hartford, CT	4.02	3.37	3.38	3.09	4.00	3.94	3.51
Wilmington, DE	3.39	3.23	3.57	3.90	4.03	3.59	2.89
Atlanta, GA	4.43	4.02	3.41	4.73	3.41	3.17	2.53
Grand Rapids, MI	3.56	3.03	3.86	3.02	3.45	3.14	2.89
Lexington, KY	4.01	4.23	4.25	4.95	4.96	3.28	2.26

Summertime Rainfall Eastern USA (in inches)

Location	Apr.	May	June	July	Aug.	Sept.	Oct.
Minneapolis, MN	2.03	3.20	4.07	3.51	3.64	2.50	1.85
Des Moines, IA	3.21	3.96	4.18	3.22	4.11	3.09	2.16
Kansas City, Mo	2.68	3.42	4.13	3.49	3.16	3.33	2.54
Little Rock, AR	5.41	5.29	3.67	3.63	3.07	4.26	2.84
Dallas, TX	3.63	4.27	2.59	2.00	1.76	3.31	2.47
Omaha, NE	2.94	4.33	4.08	3.62	4.10	3.50	2.09

During the summer growing season soil moisture loss in most of the Eastern USA can be as great or greater than it is in Cascadia. But summer rainfall is much greater. The amount of moisture lost from cultivated soils depends a great deal upon the nature of the vegetation covering the land. Bare land does not lose nearly as much water as it does when the crop has formed a thick layer of leaves that shade the earth. More will be said about this soon. Because of this variability farmers who irrigate use "evaporation from reservoirs" as a rough guide to help them know how much water to add. Weekly statistics are often printed in farm papers.

Evaporation from Reservoirs (inches per month)

Location	Apr.	May	June	July	Aug.	Sept.	Oct
Seattle, WA	2.1	2.7	3.4	3.9	3.4	2.6	1.6
Baker, OR	2.5	3.4	4.4	6.9	7.3	4.9	2.9
Sacramento, CA	3.6	5.0	7.1	8.9	8.6	7.1	4.8
Macon, GA	4.3	5.1	6.2	6.3	5.8	5.2	4.2
Minneapolis, MN	1.7	3.2	4.4	6.0	5.8	4.6	3.0
Kansas City, MO	3.1	4.4	6.1	8.0	7.8	6.0	4.5
Nashville, TN	3.3	4.1	5.1	5.8	5.4	4.9	3.7
San Antonio, TX	5.6	6.5	8.4	9.4	9.4	7.8	6.8

From May through September a reservoir near Seattle usually loses 16 inches of water by evaporation. Baker, Oregon is located east of the Cascade Mountains. Summer there is quite hot; humidity is low. Sacramento, California is even hotter and drier. The table above clearly shows that in almost every part of the United States summertime moisture loss exceeds the amount received from rainfall. Cascadia, where I lived, the loss is far greater than summer rains. In most of the humid eastern states this loss slightly exceeds the usual amount of rainfall. But normally there is quite a bit of water already present in the soil when the gardening season begins. By creatively using and conserving this moisture, gardeners with deep soil that holds a fair bit of water at the beginning of the growing season can go through an entire summer without irrigating. Gardeners with less than ideal soil conditions can still greatly reduce their dependence on irrigation.

In the Deep South summer is the time of the greatest rainfall. Still, there can be many weeks between significant rain events. Gardeners who plan for this can greatly reduce or eliminate their need to irrigate.

All the tables in this chapter were excerpted from *The Water Encyclopedia* by Frits van der Leeden and others, Lewis Publishers. It is available free online from several sources.

Chapter 2

Water-Wise Science

Plants are Water

When a living, growing plant is cut and fully dried it loses most of its weight. Most of a living plant is water. Plants conduct almost all their chemical processes in a water solution. Nearly every substance that plants move through their tissues is dissolved in water. When insoluble starches and oils are mobilized for plant energy, enzymes change them into water-soluble sugars. Even cellulose and lignin, which are insoluble structural materials that plants cannot convert back into soluble materials, are made from molecules that once were in solution.

Water is so essential that when a plant can no longer draw as much water from the soil as it is losing to the atmosphere it wilts. This is because drooping leaves lose less moisture as the sun glances off them. Some weeds can wilt temporarily and resume vigorous growth as soon as their water balance is restored. But most vegetable species aren't that resilient—vegetables may survive an afternoon of temporary wilting, but once stressed this way both the quantity and quality of their yield usually drops markedly.

Soil's Water-holding Capacity

I already mentioned that in much of the United States most vegetable species may be successfully grown with very little or no supplementary irrigation because they're capable of accessing water already stored in the soil. A smart gardener fully understands how soil moisture behaves and manages it to the benefit of his crops.

Soil mineral particles hold on to moisture using a physical property called adhesion. Here is an example of adhesion: I'm sure that at one time or another you have picked up a wet stone from a river or by the sea. A thin film of water clings to its surface. This is an example of adhesion. The more surface area there is, the greater the amount of moisture that can be held by adhesion. If we crushed that stone into dust, we would greatly increase the amount of water that could adhere to the original material. Clay particles are so small that clay's ability to hold water is not as great as its mathematically computed surface area would indicate.

Surface Area of One Gram of Soil Particles[1]

Particle type	Diameter in mm	Particles/gm	Surface (cm^2)
Very coarse sand	2.00–1.00	90	11
Coarse sand	1.00–0.50	720	23
Medium sand	0.50–0.25	5,700	45
Fine sand	0.25–0.10	46,000	91
Very fine sand	0.10–0.05	772,000	227
Silt	0.05–0.002	5,776,000	454
Clay	Below 0.002	90,260,853,000	8,000,000

Comprehending the relationship between particle size, surface area, and water-holding capacity is so essential to understanding plant growth that the surface areas presented by various sizes of soil particles have been calculated. Soils are rarely composed of a single size of particle. If the mixture is primarily sand of various sizes, we call it a sandy soil. If the mix is primarily clay, we call it a clay soil. If the soil is a mixture of sand, silt and clay, containing no more than 35 percent clay, we call it a loam. If sand predominates in the loam it is called a sandy loam. There are silt loams, clay loams and silty clay loams.

---

[1] Foth, Henry D., *Fundamentals of Soil Science,* 8th ed. (New York: John Wylie & Sons, 1990).

Available Moisture (inches of water per foot of soil)[2]

Soil Texture	Average Amount
Very coarse sand	0.5
Coarse sand	0.7
Sandy	1.0
Sandy loam	1.4
Loam	2.0
Clay loam	2.3
Silty clay	2.5
Clay	2.7

Adhering water films vary in thickness. If the coating of water adhering to a soil particle becomes too thick, the force of adhesion becomes too weak to resist the force of gravity; consequently some water flows deeper into the soil. When water films are relatively thick the soil feels wet and plant roots can easily absorb moisture. The term *field capacity* describes soil particles holding all the water they can against the force of gravity.

The thinner the water films become, the more tightly they adhere and the drier the earth feels. At some degree of desiccation, roots can not withdraw moisture from soil as fast as the plants are transpiring it. This condition is called the "temporary wilting point". The term "available moisture" refers to the difference between field capacity and the amount of moisture left after the plants have wilted and died.

Clayey soil provides far more available water than sand. It might seem logical to conclude that a clayey garden would be the most drought resistant. But there's more to it. A clayey loam can provide just about as much usable moisture as clay. Loam soils usually allow extensive root development, so a plant with a naturally aggressive and deep root system

---

[2] The term "available moisture" means the amount of water plants can extract from soil *that starts* out at full moisture capacity *until* the plants wilt temporarily. This chart makes no allowance for the amount of soil organic matter. Decomposing organic matter can hold on to and release much more moisture than mineral particles. Source: Fundamentals of Soil Science.

may be able to occupy a much larger volume of loam, ultimately coming up with more moisture than it could obtain from a heavy, airless clay. And loam topsoils often have a clayey, moisture-rich subsoil.

Because of this interplay of factors, how much available water your garden soil can provide can only be discovered by experience.

How Soil Loses Water

Imagine that we kept a test plot entirely bare all summer by rigorously hoeing it as often as any weeds appeared and hoed once every three weeks no matter what. Because plants growing around the edge of this plot might extend roots into it and extract moisture, we'll make our imaginary tilled area 50 feet by 50 feet and make all moisture measurements in the center. And let's locate this imaginary plot in full sun on flat, uniform soil.

Let's also suppose there has been plenty of rain and/or snow during winter so that on April 1 the soil is holding all the moisture it can hold as far down as there is soil. From early April until well into September the hot sun will beat down on this bare plot. Let's also suppose this plot is located in Western Oregon, where summer rains are very rare and generally come in insignificant installments that do not penetrate deeply; any rain that does fall quickly evaporates from the surface inch without recharging deeper layers. Most people would reason that a soil moisture measurement taken 6 inches down on September 15th should show very little water left. In fact, most gardeners would expect that there would be very little water found in the soil until we got down quite a few feet—if there were several feet of soil.

But that is not what happens! The hot sun does dry out the surface of this imaginary test plot but if on September 15th we dig down 6 inches or so we would encounter damp soil. Bare *loose* earth does not evaporate much water. Once a thin surface layer has been completely desiccated very little loss of moisture occurs further down. The only soils that dry out rapidly when bare are certain kinds of very heavy clays that form deep cracks as they dry out. These ever-deepening openings allow the air to evaporate additional moisture. But if the cracks are filled with dust by surface cultivation even this soil type ceases to lose water.

Soil serves as our moisture bank account, holding available water in storage. Hot sun and wind working directly on soil don't *evaporate* very much water; most moisture loss is caused by hot sun and wind working on plant leaves, making them *transpire* moisture drawn from the earth through their root systems. Plants will desiccate soil to the ultimate depth and lateral extent of their rooting ability, and then some. The extent and depth of vegetable root systems is greater than most gardeners would think. The amount of soil moisture potentially available to sustain vegetable growth is also greater than most gardeners think. The scientific name for the total of both kinds of moisture loss is evapotranspiration—evaporation from the soil's surface plus transpiration from plants.

Rain, snowmelt and irrigation are not the only ways soil moisture is replenished. If the soil is deeper than plant root systems can penetrate, water will gradually rise from below the root zone by a process called capillary flow. Capillarity works by the very same force of adhesion that makes moisture stick to a soil particle. A column of water in a vertical tube (like a thin straw) adheres to the tube's inner surface. Adhesion lifts the edges of the column of water. As the tube's diameter becomes smaller the amount of lift becomes greater. Tall trees lift moisture to their top leaves using this principle. The height limit of trees is the maximum height water can be lifted by capillarity in a microtube against the force of gravity. Soil particles form interconnected pores that act like tubes, allowing an inefficient capillary flow, recharging dry soil above from moist soil below. However, the drier soil becomes, the less effective capillary flow becomes. That is why just an inch of loose, thoroughly desiccated surface layer acts like a powerful mulch.

Industrial farming (and modern gardening) discounts the replacement of surface moisture by capillarity, considering this flow to be insignificant compared with the moisture needs of crops. But conventional agriculture these days has become a lot like intensive gardening. Farmers focus on obtaining maximum yields through establishing the highest possible plant density on heavily fertilized soil. Capillary uplift is too slow to support dense crop stands. Yet when a single plant occupies a large enough area without any competition, moisture replacement by capillarity becomes useful.

Gardeners know that plants acquire water and nutrients through their root systems and leave it at that. But the process is not that simple. Actively growing tender root tips and almost microscopic root hairs close to the tip collect most of the plant's moisture *and nutrients*. As the root grows, parts of it behind the tip cease to be effective because as the root extends it also thickens and develops a skin, while the older absorbent hairs slough off as new ones form at the ever-extending tip. This rotation from being actively foraging tissue to becoming more passive conductive and supportive tissue is an example of intelligent design because capillary uplift fails to replace soil moisture as fast as the plant might like. The plant is far better off aggressively seeking new water and nutrients in unoccupied soil than waiting for the soil that its roots already occupy to be replenished.

Lowered Plant Density: The Key to Water-Wise Gardening

I learn constantly; consequently my newer books sometimes contradict my earlier efforts. *Growing Vegetables West of the Cascades* (1989 edition) recommended somewhat wider interplant spacings on raised beds than I did in the first edition (1980) because I'd repeatedly noticed that once a leaf canopy forms, plant growth slows markedly and the amount of fruit being set also drops off greatly. Adding more fertilizer and irrigating more frequently helps after plants "bump," but still the rate of growth and production of new fruit never equals that of younger plants. For years I incorrectly believed that plants stopped producing as much after their leaves completely shaded the earth because of light competition. But now I understand that their root systems expand as fast as their above-ground parts. Unseen competition for root room slows them down as much or more than light competition does. Even if moisture is regularly recharged by irrigation and even if depleted nutrients are replaced, once a bit of earth has been occupied by the roots of one plant it is not readily available to the roots of another. So allocating more elbow room allows vegetables to grow larger, yield longer and allows the gardener to reduce the frequency of irrigation.

Withdrawals of moisture from great depths are made by growing plants. The amount of water a growing crop transpires is determined first by the nature of the species itself, by the amount of leaf surface, by the intensity of sunlight, by the air temperature, humidity, and wind speed. In these respects, the crop is like an automobile radiator. With cars, the more metal surfaces, the colder the ambient air and the higher the wind speed, the better the radiator can cool. In the garden, the more leaf surfaces, the faster, warmer, and drier the wind, and the brighter the sunlight, the more water is lost through transpiration. If there are no plants and the bare surface inch is kept loose so it dries out nearly completely, most soil water will stay unused through the entire growing season. Once a crop canopy is established the rate of water loss under it will be no less than the amount that evaporates from a reservoir—and it could be more than that.

Plants must use some of their precious energy in order to extract water from soil. The correct name for this effort is "moisture stress". Any degree of moisture stress slows growth to the same degree because it uses up energy the plant could have used to form new roots or leaves. When the soil is quite damp moisture stress is inconsequential. Growth slows noticeably after available soil moisture has lessened about half way from capacity to the wilting point. Even greater moisture stress may not be enough to cause temporary wilting but it hugely slows growth. On very closely planted beds in hot sunny weather a crop can experience strong moisture stress within a day or two of being irrigated. But if that same crop were planted far less densely it might grow well for a few weeks without irrigation. And if that crop were planted even farther apart so that no crop canopy ever developed and a considerable amount of bare, dry earth were showing, this apparent waste of growing space would result in an even slower rate of soil moisture depletion. On deep, open

soil the crop might grow on through many rainless weeks and still yield a
respectable amount without needing any irrigation at all. This paragraph
has summed up the essence of dry gardening.

You might not be so fortunate as to be gardening in deep, open,
moisture-retentive soil, but all gardeners except those with the shallowest
soil can increase their use of the free moisture nature provides and
also lengthen the time between irrigations. Most gardens can yield
abundantly if only we reduce competition for available soil moisture,
judiciously fertigate some vegetable species, and practice a few water-wise
techniques.

Does lowering plant density equally lower the yield? Surprisingly, the
amount harvested does not drop proportionately. In most cases, having
a plant density one-eighth of that recommended by intensive gardening
advocates will result in a yield about half as great as on closely planted
raised beds. Reducing plant density by half does not reduce yield by half.
What a lower plant population does achieve is to encourage all the plants
to grow larger and faster. The size of the carrot or head of broccoli will
increase. However, it may take a few days longer for that larger head or
root to develop. In my experience what gets harvested will taste and look
better.

Dealing With a Surprise Water Shortage

Suppose you already are growing an irrigated garden and something
unanticipated reduces or eliminates your water supply. Perhaps you are
homesteading and your well begins to dry up. Perhaps you're a backyard
gardener and the municipality restricts water usage. What to do?

One of the smartest things any gardener should have done long before
such a situation arises was to dig a test hole to discover the depth and
nature of their soil and thereby estimate how much water it can hold. If
your garden is on deep moisture-retentive soil and it is possible to irrigate
it before the restrictions take effect, water the entire area very heavily to
ensure there is maximum subsoil moisture. Try to saturate the soil three
feet deep, or even deeper. To guesstimate the *minimum amount* of water

to add, consult the table "Amount of water needed to bring 12 inches of soil from 70 percent of capacity to full capacity" in the last chapter of this book.

The intensive method encourages gardeners to transplant seedlings between rows of plants that are approaching harvest. If you already have done that, remove these seedlings. Then ruthlessly pull out at least half of all plants that are at two or more weeks away from harvest. For example, suppose you've got a four foot wide intensive bed holding seven rows of broccoli on 12-inch centers, say 21 plants. Remove three of those seven rows, leaving four rows that are 24 inches apart. Each row holds three plants. Also remove the plant in the center of the bed so there are two plants in each of the four remaining rows, or eight plants where there were 21 before. Then shallowly hoe the soil every two days to encourage the surface inch to fully dry out and form a dust mulch. You are now dry gardening! Now start fertigating if you can.

Chapter 3

Minimizing Irrigation

When I lived in Western Oregon I could count on starting the growing season with the full depth of our soil holding all the moisture it was capable of retaining. This blessing is common in most parts of the United States located east of the 98th meridian. However, moving west of the 98th meridian makes this blessing become increasingly unlikely, and makes it increasingly likely that irrigation will be essential. The last part of this book discusses how to make frugal and effective use of irrigation.

There are a number of ways to better sustain summertime vegetables on naturally occurring soil moisture. The most obvious step is thorough weeding starting early in spring and continuing through the entire growing season. As soon as the surface inch starts drying out we can keep it fluffed up with a hoe or with a rototiller. This breaks the surface's capillary connection with deeper soil and accelerates the formation of a dust mulch. Ongoing elimination of weeds by hoeing instead of hand-pulling helps to maintain the dust mulch. If it should rain during summer we can hoe or rotary cultivate a day or two later to restore the dust mulch. We can also spread straw or other mulch late in spring *after* the soil has warmed up. This need not be a weed-stopping thick mulch (see *Gardening Without Work* by Ruth Stout). Grain straw only half an inch thick will do it. If the mulch is not too thick it remains possible to slide an ordinary hoe under the mulch to control the weeds that do emerge through it.

Sadly, I must warn you that mulching with hay or straw these days is not safe and trouble-free like it was in Ruth Stout's day. These days farmers use highly persistent herbicides such as picloram, clopyralid, and aminopyralid to kill broadleaf weeds. Your neighbor's grass clippings may also contain these poisons. These chemicals contaminate materials gardeners then use for mulch; they also contaminate horse and cow manure. They destroy a garden; their effect can persist for several years after the manure was dug in or after mulch has been raked up and removed. So be cautious.

Encouraging Bigger Root Systems

Plants obtain much of their water supply and nutrition by expanding their root system into as yet unoccupied moist earth. Encouraging rapid root growth is the key to having a fast-growing healthy garden. Root cells must breathe oxygen. This is obtained from air held in spaces between soil particles. Soil dwellers ranging in size from bacteria to moles also use this oxygen. A slow exchange of gases does occur between soil air and free atmosphere, but in the soil there will inevitably be less oxygen than in the atmosphere and the deeper one goes the less oxygen there will be. Different plant species have varying degrees of tolerance for lack of oxygen in their root zone but they all stop growing at some depth, which limits their ability to extract moisture.

Soil compaction reduces the overall air supply and slows the exchange of soil air. Compacted soil also restricts root system expansion. When gardening with unlimited irrigation or where rain falls frequently, it is quite possible to produce what seems like satisfactory growth (to someone who has not seen what truly healthy plants can do) when only the surface 6 or 7 inches of soil facilitates root development. When gardening with limited water, China's the limit, because if soil conditions permit, many vegetable species are capable of reaching 4, 5, and even 8 feet down to find moisture and nutrition.

Evaluating Potential Root Zone

It is instructive to rent or borrow a hand-operated fence post auger and bore a three foot deep hole. It can be even more educational to extend the auger's shaft and try to bore down another 2 or 3 feet. In soil that is free of stones, using an auger is more instructive than using a conventional posthole digger or digging a small pit with a spade because where soil is naturally loose, the auger penetrates rapidly. But when the soil is

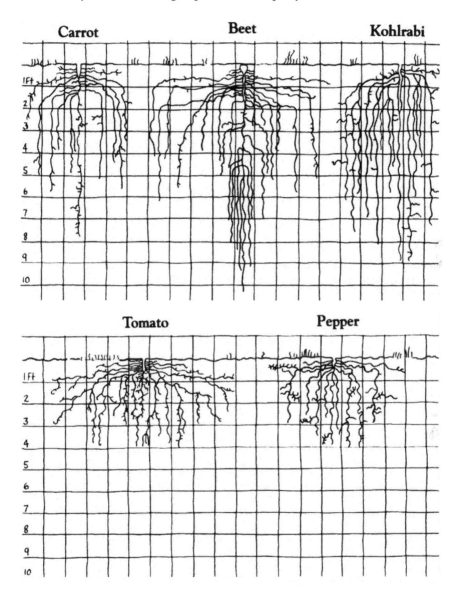

Each square in the grid is 12 inches by 12 inches.

compacted the auger turns without much effect. If your soil is stony the
more usual fence-post digger or common spade works much better.

If you find four feet of soil that is not just sand then the site provides
great dry-gardening potential that increases with the additional depth.
Some soils, especially those on the floodplains of rivers, can be over 20
feet deep and hold far more water than the deepest roots could draw
upon. Gently sloping land sometimes is covered with 5 to 7 feet of open
usable soil. However, soils on hillsides become increasingly thin with
increasing slope.

Whether an urban, suburban, or rural gardener, you should make no
assumptions about the depth and openness of the soil at your disposal.
Dig a test hole. If you find less than 2 unfortunate feet of earth before
hitting bedrock or gravel, not much water storage can occur. And if that
be the case, you will find the most helpful part of this book to be the
section about sprinkler irrigation.

Eliminating Plowpan

Deep though the soil may be, any restriction to downward root expansion greatly limits the ability of plants to aggressively find water. Topsoil
close to rivers or on floodplains may be loose and open but it may
cover a layer of droughty gravel that effectively stops root penetration.
A compacted subsoil or even a thin compressed layer such as plowpan
may also function as a barrier. Though moisture will still rise slowly by
capillarity and recharge soil above plowpan, plants obtain much more
water by rooting into unoccupied, damp soil.

Plowpan is often encountered when gardening on soil that was once
farmed. A plowpan may be found in suburbia where a farm was
converted into building blocks. Traditionally, American croplands have
been fitted for sowing with the moldboard plow. As this implement
first cuts and then flips over a 6- or 7-inch-deep slice of soil, the
plow sole—the part supporting the plow's weight—presses heavily on
the earth about 7 inches below the surface. With each subsequent

plowing the plow's weight rides at the same 7-inch depth and a compacted layer develops. Once formed, plowpan prevents the crop from rooting into the subsoil. Since winter rains leach nutrients from the topsoil and deposit them in the subsoil, plowpan prevents access to these nutrients and effectively impoverishes the field. Louis Bromfield, an American writer, returned from Paris in the late 1930s, bought a non-productive farm in Ohio and set out to restore it. His books *Pleasant Valley* and *Malabar Farm* tell the story of how he defeated a plowpan and allowed his crops to discover a rich storehouse of subsoil nutrients. They also happen to be a great read; I highly recommend them.

A plowpan also acts like the plug in a drain pipe; it slows the infiltration of water, so when it rains long and hard if the land has any slope to it a lot of water will run off the surface instead of saturating the subsoil. The last thing you want is to allow useful rainfall to escape. If there is no slope then long-lasting puddles form. So every few years the smart farmer who uses a moldboard plow also uses a subsoil plow to fracture the pan.

Plowpan can seem as firm as a rammed-earth house; once established, it can last a long, long time. My own garden along the Umpqua River in Oregon was once a wheat farm. From about 1860 through the 1930s, the field produced small grains. After wheat became unprofitable, probably because of changing market conditions and soil exhaustion, the field became an unplowed pasture. Then in the 1970s it grew flower bulbs, occasioning more plowing. All through the eighties my soil again rested under grass. In 1987, when I began using the land, there was a 2-inch-thick, very hard layer starting about 7 inches down. Below 9 inches the open earth was soft as butter extending as far down as I've ever dug.

On a garden-sized plot, plowpan is easily fractured with a spade, with a spading fork or with a very sharp combination shovel. After normal rotary tilling, which only reaches 5 to 6 inches down, either tool can fairly easily be wiggled 12 inches into the earth and small bites of plowpan loosened. Once this laborious chore is accomplished the first time, deep digging will be far easier— and probably never again necessary.

Curing Clay

The topsoil may seem very open and friable but how about the subsoil? Over geologic time, mineral particles are slowly dissolved by weak soil acids; clay forms from recombination of the breakdown products. Then rainwater flowing through the soil transports these minuscule clay particles deeper, where they concentrate. In humid climates it is the usual thing to find a foot or two of loamy topsoil with dense clay beneath it. The soil scientist calls this situation a duplex soil. Machine tillage does not improve a clay subsoil. Temporarily stripping away the topsoil, then painstaking double or even triple digging compost, lime and possibly gypsum into the clay as deeply as possible (according to what a proper soil test says is needed) will loosen the clay. After the topsoil has been put back in place the crops will grow much better than they otherwise might have but I do not think there are many people who could be motivated to do this much work. In case you are highly motivated and possess enormous physical energy, John Jeavons's book *How To Grow More Vegetables* details this double digging procedure.

I have done soil testing for gardeners with fine sand topsoil that contain a lot of iron. A foot down lies what seems to be hard, almost rocklike yellow-red "clay". Gardeners who have shown me this "clay" are surprised when I tell them it is not clay at all, but is the same sand found in the topsoil, but iron has been leached out of the surface and deposited there. The iron acts like cement. Anyone willing to do enough hard work can pickaxe their way into this hard stuff, mix compost into it and turn it into root zone. In this situation it would also be smart to read my book *The Intelligent Gardener,* do a soil test and when mechanically breaking it up, mix in soil amendments that bring calcium and magnesium into balance.

As I write this section I find myself wishing I could explain a large chunk of *The Intelligent Gardener* in just a few paragraphs so it would fit into this book. I apologize: I can not. The best I can say in-brief-made-simple is to strongly advise those who garden clay or have duplex soils with clay subsoil, to avoid dolomite lime as though it were soil poison (unless you know better from a proper soil test) because the amount of magnesium in clay strongly influences how compacted and airless it can be, how open and crumbly it can become. What you want to achieve with clay is to have it contain as much calcium as possible without pushing pH up too high and not contain too much magnesium. If I had to improve clay without benefit of a full laboratory soil test I would blend as much high calcium aglime into the surface six inches as the soil could accept without pushing soil pH over 6.5 (never ever more than 4 tons of aglime per acre) and then regularly spread small doses of aglime, say 1 ton of lime per acre per year. Slowly slowly slowly this aglime dissolves; rain and irrigation moves calcium deeper. When clay holds enough calcium it *flocculates,* meaning it gets crumbly and holds more air.

An impenetrable clay subsoil can still supply crops with surprisingly large amounts of moisture. It will do this even better if it becomes more accessible to root penetration. As rooting depth increases the organic matter content and accessibility of this clayey layer can be further improved through intelligent green manuring and liming the topsoil. Green manuring will soon be discussed.

Not all clay soils are compacted and airless. For example, on the gentler slopes of the Cascades foothills and other places throughout the maritime Northwest whose soil formed out of the basalt flows which form the base of the Cascade range, whose soil developed out of a deep, friable, red clay called (in Oregon) Jory. This clay can be 6 to 8 feet deep and is sufficiently porous and well drained to use for highly productive orchards. Water-wise vegetable gardeners can do wonders with Jory and other similar soils, though clays never grow the most perfectly shaped carrots and parsnips.

The bottom line about gardening clay is that it'll go much better if you understand the chemical nature of clay and how to improve it by balancing calcium and magnesium. To achieve that I'm trying to sell you a copy of my book *The Intelligent Gardener.* Keep in mind that every copy I sell earns me slightly more than one dollar.

Spotting a Likely Site

Wild plants can reveal a likely dry gardening site. In the western states of Oregon and Washington, Himalaya or Evergreen blackberries take over neglected pastures and other sunny clearings. Where there is not much available soil moisture they grow 2 feet tall and produce small, tasteless fruit. Where they grow 6 feet tall and the berries are sweet and good sized there is deep, open soil. When the berry vines are 8 feet tall and the fruits are especially huge, usually there is both deep, loose soil and a higher than usual amount of fertility.

For years I wondered at the sad appearance of Douglas fir trees in the vicinity of Yelm, Washington. Was that due to extreme soil infertility? Then I learned that conifers respond more to summertime soil moisture than to plant nutrients. I obtained a soil survey and discovered that much of that area was covered with gravelly coarse sand—glacial till. Eureka! I apologize that I am not able to provide similar information for other areas of North America. But I am sure there are similar phenomena in your region.

The Soil Conservation Service (SCS), a U.S. Government agency, has probably put a soil auger into your land or close by. Its tests have been correlated, mapped and categorized by texture, depth, and ability to provide available moisture. In 1987, when I was in the market for a new homestead, I first went to my county SCS office, mapped out locations where the soil was suitable, and then went hunting.

Topsoil Moisture Retention

When I advise someone starting a new garden I routinely prescribe spreading a layer of high quality compost that is one inch thick. It should be dug or rototilled into the top six inches. After that, once-a-year spread a layer one-half inch thick on the surface like a mulch, where it actually works better than were it dug in. The compost will get mixed into the soil when you hoe weeds and make seedbeds. Worms and other soil life will come to the surface to eat it and then eliminate it from their digestive system deeper in the soil.

Having sufficient soil organic matter is essential to plant health and essential to bringing the soil into a condition of crumbly workability called "good tilth". Soil given adequate organic matter does not dry out as rapidly. It digs and hoes easily. Germinating seeds can push their way through it. Seedbeds rich in organic matter do not need to be watered as often. I suggest shallowly hoeing in compost before sowing seeds because that way the surface inch behaves like potting mix.

Garden books frequently recommend adding extraordinarily large amounts of organic matter. Let's consider how much additional soil moisture might be made available by hugely increasing organic matter. Here's a fact you can depend upon: increasing Soil Organic Matter (SOM) by 1 percent in the top six inches of a 100 square feet of growing bed can provide plants with an additional 40 gallons of water. I have interpreted a great many soil tests for Tasmanian gardeners who had spread considerably larger quantities of compost than I recommend and done that over many years. In my climate their garden soil has stabilized at around 10% SOM. To put that fact in better perspective, when productive pasture in Tasmania has been recently converted to vegetable garden the soil test usually shows 5% SOM, maybe 7% if the pasture was unusually fertile. Increasing that amount to 10 percent in the top 6 inches of 100 square feet of bed means the water holding capacity has been upped by as much as 200 gallons.

Trying to build up soil organic matter over 10 percent is self-defeating because the higher the level gets, the more nitrogen gets released and the more rapidly organic matter decays. Obtaining and spreading enough compost to cover the garden one inch deep once a year, every year for quite a few years is what it takes to build SOM to around 10 percent in a climate where summer is not too hot—like most of Cascadia or the northern tier of eastern states. Where summer is hot bringing in that much compost every year would only lift SOM to 4 percent.

The higher the soil temperature, the faster SOM decays and is lost. Hans Jenny's monumental book *Factors In Soil Formation** says that if SOM were measured in pastures located near the Mississippi River, in steamy Arkansas there would be around 2% SOM, maybe 2.5%, while in Southern Wisconsin where summer temperatures are more moderate it might be 4% to 4.5%. In Northern Wisconsin it might be 5%. In the cooler parts of Cascadia it might reach 7%. Achieving SOM of 4% in Arkansas would be as difficult as reaching 10% around Portland, Oregon.

Some gardeners think it is possible to massively increase soil organic matter by growing green manures and tilling them in. This is true if you leave out the word "massively". Upping soil organic matter by 1 percent per year is possible with green manures—for the first few years. But after that the rate of increase slows greatly and then the level reaches a peak that can not be exceeded.

I hope you now understand there is a limit to how much more moisture soil can be induced to hold. Increasing organic matter is helpful. It'll get a garden through longer rainless spells without needing to irrigate. But we're only talking about a week longer or maybe two weeks longer.

Another important benefit from building higher levels of topsoil organic matter is that the infiltration of rain is enhanced. There is drizzle, there is rain and then sometimes there is a downpour. I once had to pull to the side of the road during an afternoon thunderstorm in Tennessee because I could not see the front end of my own car nor any sign I had the headlights on. This kind of rain event is normal in much of the humid eastern states. If your garden slopes more than 1 foot of drop in 100 feet of run (1%) then water might run off downhill instead of percolating into the earth when rain buckets down. When gardening without irrigation it is essential that every drop of rain or spring snowmelt goes into the soil and not run off. An open spongy topsoil achieves that.

Smart gardeners who must grow vegetables on a sloping site will arrange their beds to follow contour lines. That way the beds and paths between them have no slope; they look like stair steps. Temporary puddles may form on compacted paths but the water will not flow away; it will soak in. If I am describing your situation I urge you to spend time observing the garden with the idea of stopping runoff, keeping it on the site so it percolates into the soil rather than being lost.

Improving Subsoil, More About

Decaying roots form long-lasting passageways that facilitate air exchange and reduce run off during heavy rains. A healthy green manure also puts organic matter into the subsoil, which increases its ability to store moisture. Green manuring is a complex subject that deserves a very large book of its own. Choice of an appropriate species, when to sow it, how thickly to sow the seed, how deeply to position it and so forth differ by region. I am only qualified to write about Cascadia; non-Cascadians will have to expand upon the few generalities I can provide. Green manuring is routine farming practice. You will find specific detailed information from your own state's extension service.

Green manuring can be risky. The largest danger comes when doing it on clayey- or poorly-drained land that does not allow tillage when wet. As the farmer would describe this catastrophe—the crop may get too "forward" before it can be plowed down. When green manure crops begin forming seed the vegetation becomes woody, which makes it slow to decompose. If the green manure is turned under before flowering has barely begun it will rot within a few weeks. The following crop grows well. But once seed stalks have noticeably emerged from the base of grasses or grass-cereals, when flowers start forming on clovers and etc, it may take six to eight weeks after being plowed down before the material breaks down enough to allow the next crop to grow well. Because of all these concerns I strongly suggest the novice do trial plots using various species before expanding it to all or most of the garden.

In Cascadia it is routine to overwinter green manures, usually sowing them early- to mid-October, which is before the soil gets too cold for germination to occur; turning them under in spring before the crop exhausts subsoil moisture and before the crop starts getting woody. Establishing the green manure crop requires getting the summer vegetables out of the way in time for the short window of opportunity that allows sowing it. When irrigation is not possible getting green manure seeds to sprout requires meaningful rain before mid-October, which can not be counted upon. For Cascadia, I recommend crimson clover, tic beans (small seeded favas) and cereal rye. I also recommend these two OSU publications:

https://extension.oregonstate.edu/crop-production/vegetables/cover-crop-establishment-western-oregon-vegetables

https://extension.oregonstate.edu/crop-production/soil/establishing-winter-cover-crops

In Cascadia I most frequently used crimson clover. It can grow mid-thigh high before flowering starts in April. If turned under when flowers first begin to form the nitrogen-rich vegetation decomposes rapidly, leaving a beautiful seedbed. If it is turned under after the plants are blooming the vegetation will tangle tiller tines and be difficult to manage. When an unusually rainy spring prevented me from turning crimson clover under in time I had to mow it close to the ground, rake up the vegetation and use it to make compost. Only then could I till in the stubble and make a seedbed.

Cereal rye is a particularly useful green manure. It is also cold hardy enough to overwinter under the snow in much of the United States. Establish it early in autumn. In Cascadia it puts up seed stalks mid- late April. A few days after the seed heads have pollinated mow them close to the ground and leave the straw where it fell, such that it dries out in a few days and serves as a thin mulch. I mow with a weedwacker/brushcutter. *Waiting until pollination has happened before mowing is essential to success.* If rye is mowed before pollination the plants resprout from the base, turning the green manure into hard-to-eliminate moisture-draining weeds. But after pollination cereal rye plants die immediately after being mowed close to the soil. The root system of cereal rye is extremely dense, very fine textured (and deep); it rots rapidly, naturally fracturing the soil so it will not need further preparation to grow vegetables. No tillage will be necessary other than making small holes in which to set transplants or narrow rows loosened with a hoe for sowing seeds. I broadcast fertilizer into the standing crop just before mowing it or else into the drying straw soon after mowing.

Green manures transpire a great deal of subsoil moisture if they grow large before they are turned under. This moisture could have supported summertime vegetable crops. If your district gets considerable spring rainfall, clearing a green manure crop a few weeks before the last usual frost date in spring can work fine. I regret that I can't provide detailed advice because I have no hands-on experience in your region.

For gardeners with excess land, the best possible soil-building crop rotation includes several years of perennial grass-legume-herb mixtures to maintain the openness of the subsoil followed by a few years of vegetables and then back to pasture (see Frank Newman Turner's books listed in More Reading). I gardened this way when I lived on 16 acres of deep alluvial soil in Cascadia. In October, after a few inches of rain had softened the earth, I'd spread 50 pounds of agricultural lime per 1,000 square feet and break the thick pasture sod covering next year's garden plot by shallow rotary tilling. The sod only partly decomposed over the winter and a great deal of weed/grass seed germinated—which is desirable because it made next year's garden a lot less weedy. Also, I did not have bare soil being pounded flat by heavy rain all winter. Early the next spring I'd broadcast a concoction I call "complete organic fertilizer" (see page 35 for the recipe and full directions for use) and rototilled after the soil dried down a bit. This eliminated all the winter weeds before they could begin to make seed.

For the next two years vegetables grew vigorously on this new ground supported only with a complete organic fertilizer. I made no compost nor did I import any. All crop residues were turned under with a walk-behind self-propelled rototiller. But vegetable gardening makes soil organic levels decline and favors the development of insect and disease problems. So every two years I'd start a new garden on another plot and replant the old garden to pasture. The pasture should grow for at least four years before being put back into vegetables. I never removed any vegetation during the long rebuilding under grass; I merely mowed it once or twice a year and allowed the organic matter content of the soil to increase. If there ever were a place where chemical fertilizers might be appropriate around a food garden, it would be to affordably enhance the growth of biomass during green manuring.

Were I a city gardener I'd consider growing vegetables in the front yard for a few years and then switching to the back yard. When rotating out of vegetables I would not plant lawn grasses, but instead, vigorous pasture species. In order to maximally increase soil organic matter do not mow short as you would a lawn. Ideally, the mower could be set at four inches. If possible let the grasses grow until they start putting up seed heads, then mow to four inches tall. This procedure may have to be repeated if there is much summer rainfall.

Gardeners in the Deep South could overwinter green manures as I did in Cascadia but they also have the option of starting cool season vegetables at the end of summer.

Mulching

Gardening under a permanent thick straw or hay mulch was popularized more than half a century ago by Ruth Stout and her disciples (Ruth's book *Gardening Without Work* is listed in More Reading). She asserts that mulching is a surefire way to drought-proof gardens while eliminating virtually any need for tillage, weeding, and fertilizing. I'm sure Ruth told us her truth. However, I twice attempted Ruth's method; once in Southern California and again in Western Oregon—with disastrous results in both locations.

Mulching with vegetation actually does not reduce summertime moisture loss much better than mulching with loose dry soil, sometimes called "dust mulching". However, an organic mulch can make summertime vegetables grow better than a dust mulch. Most of the organic matter in soil, and consequently most of the available nitrogen, is found in the first few inches. Levels of other available mineral nutrients are usually higher in the top six inches, sometimes called the furrowslice. However, once the furrowslice dries out no root activity will occur there and the plants are forced to feed in soil that is far less fertile. Keeping the topsoil damp especially improves the growth of predominantly surface-feeding vegetables like lettuce, onion, radish and most kinds of small seedlings. Mulch also keeps the temperature in the surface few inches in the comfort zone for plant roots and the soil biota when hot summer sun beats down.

Ruth's book says to keep the mulch in place year-round. She says mulch should be thick enough to prevent weeds from coming up through it. Anywhere a weed does appear reveals that the mulch had become too thin at that spot. The remedy is not just to pull the weed, but also to put down another flake of straw there. That worked for Ruth. I assert it worked because her garden was in Connecticut where winter is wintery.

I did most of my gardening in Cascadia. The region has long cool springs but mulch greatly slows the soil's spring warmup. Plants do not grow rapidly in cold soil. Summer nights in Cascadia are cool; the harvest of summer crops can be delayed by several weeks—even a month—when the garden is covered with mulch before the soil warms up. Many kinds of vegetables grow smaller plants in cold soil, these yield much less when they finally start to ripen fruit or form ears of corn. Worst, without a long winter freeze to set populations back, permanent thick mulch breeds a great many slugs, earwigs, and sowbugs. These primary decomposers are no problem when their numbers are few. But protected from predators by mulch, given unlimited food (the mulch) and their population not being set back by a winter that freezes soil solid, they soon become so numerous they destroy newly germinated seedlings and eat back larger leaves. In my experience a permanent mulch cannot be maintained for more than one year in California or in Cascadia before vegetable gardening becomes very difficult. I am confident this is also true in the Deep South.

Laying down mulch after the soil has warmed up well, making it no thicker than necessary to reduce evaporation, raking up what remains of that mulch in autumn or very early the next spring and composting it—all these practices prevent destructive insect population levels from developing. If the mulch overwinters in place it helpfully reduces soil compaction and enhances the survival and multiplication of earthworms. On the downside, during summer a thin mulch enhances the germination of weed seeds without being thick enough to suppress their emergence. And any mulch, even a thin one, interferes with hoeing, while hand weeding through mulch is tedious.

I've tried mulching quite a few vegetable species while dry gardening in Cascadia and found little or no improvement in plant growth with most kinds. Bare soil combined with fertigation is better all around. However, keep in mind that my experience was in a climate where summer days are warm but rarely hot, where surface soil temperatures in summer are rarely at root killing levels.

Windbreaks

Plants transpire more moisture when the sun shines, when temperatures are high, when humidity is low and when the wind blows; it is just like laundry drying on a clothesline. On a sunny 80° F day with a mild breeze and low humidity, soil that is completely shaded by a growing crop can lose 1/4 inch of water per day. On a breezy, hot and sunny day with low humidity, the amount lost can be more than double that.

A windbreak can reduce evapotranspiration to its leeward by 15% to 20%, the effect extends downwind at least 10 times the height of the windbreak. It also reduces wind speed by half out to 10 times its height. So a windbreak only three feet high can reduce moisture loss for up to 30 feet downwind. Porous windbreaks work better than solid ones; a hedge works better than a wall. In one study a 10 foot tall hedge reduced moisture loss by 31 percent downwind for 150 feet—15 times the barrier's height. For quite a few years I gardened on a windy Tasmanian hilltop. I got much better performance from heat loving vegetables by erecting a temporary windbreak made of 3 foot wide 50% shade cloth stapled to wooden garden stakes. This not only reduced stress and damage to delicate leaves (zucchini, beans), it also increased temperature by a few degrees, which makes a huge difference in the cool summer climate I live in.

Fertilizing, Foliar Feeding, and Fertigating

Perhaps the single most important way a water-wise gardener can help plants cope with moisture stress is to increase soil fertility— especially subsoil fertility. As early as 1882 it was determined that less water is required to produce dry-weight biomass when the soil is fertilized than when it is not fertilized. The experiment showed it required 1,100 pounds of water to grow 1 pound of dry matter on infertile soil, but only 575 pounds of water to produce a pound of dry matter on rich land. Poor plant nutrition increases the water cost of every pound of dry matter produced.

Fertilizers, manures, and potent composts mainly increase plant nutrients in the top six inches. But the water-wise gardener must put nutrition down deep where the soil stays damp through the summer. When the furrowslice gets dry plants may stop growing and show signs of moisture stress. Several things can be done to reduce or prevent midsummer stunting. First, before sowing or transplanting large species like tomato, squash, or big brassicas, dig a small pit 18 inches across and 12 inches deep. Into the bottom of that pit deeply dig in an inch of compost and handful of complete organic fertilizer. Also mix fertilizer and compost into the soil you just dug out before it goes back into the hole.

Complete Organic Fertilizer

Since the late 1980s my books have recommended using a one-size-fits-most organic fertilizer recipe. It works excellently in almost every food garden. I call it complete organic fertilizer, or COF. The early versions were not truly complete—but I have steadily improved it. This one is by far the best one. Repeated uses of COF gently move most soils into chemical balance *and* from the get-go it makes veggies grow fast and taste great. It entirely consists of substances that are allowed for use by organic certification agencies. Vegetables grow great when the soil is given only COF and *small* additions of compost.

COF costs only a tiny fraction of the supermarket price of the vegetables it grows—if you buy the major ingredients in 50 pound bags. Farm/ranch suppliers are the most likely sources. If you should find COF ingredients at a garden shop they will almost inevitably be offered in small quantities at shockingly high prices per pound. If I were an urban gardener I would drive to a country store and stock up.

A fertilizer that puts the highest nutritional content into the vegetables it grows must provide roughly equal amounts of nitrogen (N), phosphorus (P), and potassium (K, from kalium, its Latin name). It would also supply substantial amounts of calcium (Ca) and sulfur (S) and *effective* quantities of iodine, manganese, copper, zinc, molybdenum, boron, etc. I stress the word "effective" because a lab analysis of many prepared fertilizers (printed on the package at a garden center) may show all these elements are present but when the numbers are crunched it becomes clear that the amounts are too small to make a nutritional difference. An ideal garden fertilizer would release most of its nutrients over a few months instead of in a few days. That way they don't wash out of the topsoil with the first excessive irrigation or heavy rain. It would be odorless, finely powdered, not burn leaves if a little got on them and would not poison plants or soil life if accidentally double-dosed. All this accurately describes COF.

COF is intended to be concocted by the gardener. Precise measurement is not essential. Proportions varying plus or minus 10 percent will work out to be exact enough. I prepare mine using two 5 gallon white plastic buckets. First I put all the ingredients listed below into one bucket and then blend them thoroughly by pouring back and forth between two buckets at least 8 times. Spread the entire amount of the recipe below over 100 square feet of growing bed or intended root zone. There is no absolute need to dig or rototill it in; it also can be left on the surface or shallowly hoed in.

COF for 100 square feet

Nitrogen
5 pounds of oilseed meal, such as cottonseed meal, soybean meal, canolaseed meal; or else 15 pounds of bagged chicken manure compost.

Calcium
If you live where the native vegetation was a forest, 2 pounds agricultural lime *(not dolomite)* and 1 pound agricultural gypsum

If you live where the original vegetation was treeless grassy prairie or desert scrub, no aglime. Instead, use 2 pounds agricultural gypsum.

Phosphorus
2 pounds colloidal (soft) rock phosphate, or high phosphate guano or bone meal. If you use chicken manure compost for nitrogen then reduce the amount of phosphate fertilizer by half.

Potassium
½ pound potassium sulfate. Do not use greensand as promoted in so many organic gardening books. Do not use muriate of potash, KCl. It is slightly less costly than potassium sulfate but muriate causes leaching of calcium and degrades flavor.

Trace Elements and Micronutrients
- Borax 20 gm or ¾ ounce
- Manganese sulfate 30 gm or 1 ounce
- Zinc sulfate 15 gm or ½ ounce
- Copper sulfate 15 gm or ½ ounce
- Sodium molybdate 0.6 gm or ⅛ kitchen measuring teaspoon
- 1 pound kelp meal

Oilseed meals are a by-product of extracting vegetable oil from seeds such as soybean, linseed, sunflower, cottonseed, canolaseed (rape), etc. When chemically analyzed, most seed meals show a similar NPK content—about 5-2-1—so they must be considered a source of slow-release nitrogen that does not supply nearly enough phosphorus or potassium. Most farm soils are severely depleted of trace elements, so seed meals contain very little copper, zinc, manganese, boron, etc. Seed meals are sold as animal feed and not as fertilizer so they are labeled by their protein content. Most seed meals contain between 35 percent and 45 percent protein. The general rule is that each 6.2 percent of protein releases 1 percent nitrogen, so choose whichever type of seed meal gives you the largest amount of protein for the least cost. Seed meals will store for years if kept dry and protected from mice. Once they have been blended with the other ingredients in COF they no longer interest vermin.

There are three types of lime appropriate for gardening. "Agricultural lime" is finely ground limestone, a type of rock containing mostly calcium carbonate. Dolomite is a form of limestone containing both calcium and magnesium carbonates. Dolomite was much praised in early organic gardening literature but we know better these days. Dolomite is to be avoided on soils containing more than a very small amount of clay because high levels of magnesium make clay become sticky, airless and prone to form clods. When those same clays hold on to less magnesium and more calcium they become crumbly, less sticky, and then the soil provides plants with more air, is less prone to form clods or stick to your boots and tools. Very few soils are deficient in magnesium so when making a one-size-fits-most fertilizer recipe it seems smart to avoid dolomite. I do soil testing for many gardeners every year; rarely do I find myself prescribing dolomite or any other magnesium-rich material. Gypsum is a naturally occurring rock made of calcium sulfate. Gypsum is included because it contains sulfur, which is an essential plant nutrient. It also provides readily available calcium which often helps greatly because gypsum completely dissolves in a few months whereas most grinds of aglime take three or more years to fully dissolve into soil unless it is very fine powder (like #100). Do not use quicklime, burnt lime or hydrated lime. These are chemically very active and shockingly harsh on the soil biology.

You may have read that the soil acidity should be corrected by liming. I suggest that you forget about pH. If a soil test shows your garden needs calcium—go ahead and spread the quantity- and type of lime the soil lab recommends. The amount probably will be a lot greater than is in COF. And then use COF too. Without doing a soil test be assured that COF will provide enough calcium for plant nutrition, which is the most important thing liming accomplishes, *and* will over years of use bring pH into a desirable range.

Soft rock phosphate, bone meal, or high phosphorus guano provide what seed meal lacks. Seed meals once had quite a bit more phosphorus in them but over the 45 years I have been using them the phosphorus content has steadily decreased. Incidentally, a comparison of nutritional studies published by the USDA over the past 50 years shows that the average nutritional value of vegetables has also declined over 50 percent across the board—all vegetables, all vitamins and all minerals. This is one reason I make homegrown vegetables the largest part of my total diet. Guano, rock phosphate, bone meal, and especially kelp meal may seem costly but including them adds considerable fortitude to the plants and greatly increases the nutrient density of your vegetables. If shortage of money stops you from obtaining kelp meal or a phosphate supplement, I suggest reconsidering your priorities. In my opinion, you can't spend too much money creating maximum nutrition in your food; a dollar spent here saves several in terms of health costs of all sorts — and how do you place a money value on suffering?

Trace element fertilizers and potassium sulfate can be easily and inexpensively obtained in garden-sized quantities from alphachemicals.com. Even though these are soluble synthetic chemicals that were demonized in classic organic gardening literature, the modest but effective quantities in COF will not harm soil life. Organic certification agencies now allow their use.

Kelp meal has become shockingly expensive but one 50-pound sack will supply a large family food garden for several years. Kelp provides plants with a complete range of micronutrients such as iodine, nickel, cobalt, selenium, chromium and etc, plus natural growth regulators and hormones that act like vitamins, increasing plant resistance to stress, thereby increasing your body's resistance to stress too.

There is a way to economize with kelp. Instead of putting kelp into the soil you can foliar feed kelp extract. Foliar feeding is more time-consuming than mixing the kelp into COF, but I believe the practice has advantages beyond frugality. Since foliar feeding should be done every two weeks, the spraying of it gets the gardener walking among the plants, putting into practice the old saying that the best fertilizer of all is the feet of the gardener. Indeed, even if you can afford kelp meal, I suggest you also foliar feed kelp extract.

Using Complete Organic Fertilizer

Before planting every crop uniformly spread one full batch of COF atop each 100 square feet of raised bed or each 50 feet of planting row, in that case cover a strip 2 feet wide. Blend the fertilizer into the surface inch with a hoe, or spade it in 6 inches deep, or just spread it. Minuscule soil animals come to the surface at night and will eat it and mix it in for you after they have digested it. Hoeing weeds will mix it in too. If you're making hills for what will become large plants, mix an additional half-cup or even a full measuring cup of COF into each one.

If you want maximum results, wait a few weeks after seedlings have come up or been transplanted out then sprinkle small amounts of COF around them, thinly covering the area that the root system will grow into during the next few weeks. As the plants grow, side-dress them again every few weeks, placing each dusting farther from the plants' centers. Each side-dressing will require spreading more fertilizer than the previous one. During the time the crop is growing side-dress no more than four to six additional quarts per 100 square feet of bed or 50 feet of row. If a side-dressing fails to increase the growth rate over the next few weeks, that indicates it wasn't needed, so do it no more.

Compost

Vegetable gardening without importing compost almost inevitably depletes soil of organic matter unless much attention is paid to green manuring. In order to maintain soil organic matter (SOM) by growing it on site requires that a great part of the garden grows green manure crops, not vegetables. So gardeners make and/or buy compost. Imported organic matter provides large amounts of plant nutrients but these are usually out of balance. Depending on compost alone to provide plant nutrients usually lowers nutrient density and in cool climates does not make plants grow nearly as fast or as large as they can.

The COF recipe in this book allows for the nutrients typically supplied by a layer of compost no more than ½ inch thick, spread once a year. *Please, I entreat you,* do not spread more compost than that and do not spread less than ¼ inch per year or else the soil's organic matter level will decline. Use COF to make the plants grow big and fast, do not try to make that happen by using more compost.

There is a great deal more to know about fertilizing and maintaining organic matter than I can possibly go into in this book. For those who wish to fully understand the topic I recommend reading my book *The Intelligent Gardener,* which is available through major online booksellers.

Foliar Feeding

Foliar feeding plants when they are under moderate moisture stress keeps them growing fast. Soluble nutrients sprayed on plant leaves are rapidly absorbed, especially so when this is done early in the morning. Unfortunately, nutrient solutions that are dilute enough to not burn leaves provoke a strong growth response for only a week. In order to maintain rapid growth, foliar nutrition must be repeatedly applied. To efficiently spray a garden larger than a few hundred square feet, 1 suggest buying an agricultural-grade, 3- to 4 gallon backpack sprayer with a side-handle pump. The store that sells it (probably a farm supply store) will also support you with a complete assortment of nozzles that vary the rate of emission and the spray pattern. High-quality equipment like this outlasts many, many cheap pump sprayers designed for the consumer market, and replacement parts are also available. Keep in mind that consumer merchandise is designed to be consumed; stuff made for farming is built to last.

Using foliar fertilizers requires caution and forethought. Spinach, beet, and chard leaves seem particularly sensitive to being burnt by foliars (and by organic insecticides). Because gardens contain a broad mix of species I suggest diluting foliars at half- to two-thirds the recommended strength on the product label. Cabbage family vegetables (and some others too) protect their leaf surfaces with a waxy, moisture-retentive coating that makes sprays bead up and run off rather than stick and be absorbed. Mixing foliar solutions with spreader/sticker, with Safer's Soap, even a few drops of ordinary dishwashing liquid per quart, or, if bugs are also a problem, with a liquid organic insecticide that already contains a surfactant, allows the fertilizer to have a much bigger effect.

I am sorry to have to say this, but in terms of nutrient balance the poorest foliar sprays are the ones approved for certified organics. That's for two reasons: because it is nearly impossible to get significant quantities of phosphorus or calcium into solution using any combination of fish emulsion and brewed seaweed/kelp; and similarly, it is not possible to deliver effective concentrations of trace elements from fish or kelp. The most useful organically acceptable general purpose foliar is made by combining ½ to 1 tablespoon each of fish emulsion and liquid seaweed concentrate per gallon of water. There are many brands of liquid fish. The stuff the founders of the organic farming and gardening movement would have approved of provides mostly nitrogen with very little phosphorus or potassium. But you will find some brands in the garden center that have about as much P in them as N. This was achieved by "stabilizing" the brew with phosphoric acid, a synthetic chemical— which organic certification rabbis say is kosher. I see nothing wrong with using phosphoric acid as a nutrient for food crops but it is a synthetic material produced in a chemical vat, not naturally occurring, which those same rabbis profess to believe in. Neither fish nor kelp provide enough trace elements (copper, zinc, boron, manganese, molybdenum) to make a meaningful difference.

Foliar spraying and irrigating with fertilized water (fertigation) are two occasions when I use water-soluble chemicals. The most readily available brand at garden centers is Miracle-Gro. I reject Miracle-Gro because it is produced by a burgeoning and ravenously acquisitive global corporation. If I were limited to what's available at most garden centers I'd choose Peters 20-20-20. However, I've had the best results with Dyna-Gro, either their GROW 7-9-5 or their all purpose 7-7-7. For crops already ripening a heavy fruit load that I do not want to provoke into rapid vegetative growth, like tomato or pepper, I'd use Dyna-Gro BLOOM 3-12-6. Dyna-Gro products are not hard to source; find them online through Amazon and many other outlets.

Vegetables That:

Respond strongly to foliars			
Asparagus	Carrot	Melons	Squash
Beans	Cauliflower	Peas	Tomato
Broccoli	Brussels sprout	Cucumber	
Cabbage	Eggplant	Radish	
Kale	Rutabaga	Potato	
Have Sensitive Leaves			
Beet	Lettuce	Pepper	
Chard	Spinach		
Thrive with Fertigation			
Brussels sprout	Kale	Savoy cabbage	Broccoli
Cucumber	Melon	Squash	
Eggplant	Pepper	Tomato	

If the soil is deep and moisture retentive, generous fertigation every two to three weeks maximizes yield while minimizing water use. If the soil is sandy I'd fertigate every week to 10 days and provide half as much each time. Make the first fertigation early in summer and continue periodically until late in summer when vegetative growth slows and moisture loss drops. In the Deep South I'd fertigate year-round, but less frequently in the cool season. When I lived in Oregon I did it with a half dozen recycled plastic buckets. I fertigated 10 to 15 plants each time I worked the garden; it took no special effort to cover the entire garden every few weeks.

To make a fertigation bucket, drill one ¼ inch diameter hole through the side of a 3 gallon to 5 gallon plastic bucket; locate it ¼ inch up from the bottom. The empty bucket is positioned where fertilized water drains out as close as conveniently possible to the stem of a plant. Then the bucket is filled with water from a hose and immediately given a measured amount of liquid fertilizer concentrate. It takes a while for that much water to pass through a quarter-inch-diameter opening, and because of the slow flow rate, water penetrates deeply into the subsoil without wetting much of the surface. Each fertigation makes the plant grow very rapidly for a while. This improvement is as much the result of improved nutrition as it is from added moisture.

Ultra-conservative organic gardeners can fertigate with combinations of fish emulsion and seaweed at the same dilution the bottle suggests for foliar spraying. It is a very old practice to fertigate and foliar feed at the same time with compost/manure tea out of a watering can. Australians call this "drenching". Determining the correct strength to make compost tea is a matter of trial and error. I would choose Peters 20-20-20 mixed at half the recommended dilution or better, DynaGro mixed at what the company calls the "production rate".

When fertigating it is especially important to match the amount of water applied to the depth of the soil and to the size of the plant's root system. There's no sense adding more than it takes to saturate the plant's root zone. Calculating the optimum amount requires evaluating the following factors: soil water-holding capacity and accessible depth; how deep the root systems have developed; how broadly the water spreads out from a drip bucket and the rate of loss due to transpiration. You can assume the root system is at least the size of the above ground plant with the exception of root crops. In the case of carrot, parsnip, beetroot and parsley you can assume the root system is much deeper than the plant is tall.

I also suggest that the first time you fertigate, position a bucket where no plant is growing, put in one measured gallon of water, let it drain out, wait a few hours for everything to settle and then dig a hole there. See how deeply there is wet soil and how broadly moisture has spread out. You might also do it with three gallons and see how the treatments compare.

On sandy soil, fertigation from a bucket with one small hole in it moistens the earth nearly straight down as there is little lateral dispersion. One foot below the surface the wet area might only be 12 inches in diameter. On the surface will be a wet spot no larger in diameter than a tea mug. If water spreads wider than that on the surface then the hole in the bucket is too large; it is emitting water faster than the soil can take it in. Conversely, with a clay soil the surface may seem dry except for a small wet spot, but 18 inches away from the emission point and just 3 inches down the earth may be saturated with water, while a few inches deeper, significant dispersion may reach out nearly 24 inches.

Another important concern relates to how much fertigation water to put into a bucket. 1 cubic foot of water equals 5 gallons. A 12-inch-diameter circle covers 0.75 square feet, so 1 cubic foot of water dispersed from a single emitter into sandy soil acts like 16 inches of rainfall, hugely overwatering a medium that can hold only an inch or so of available water per foot. On heavy clay, the fertigation bucket leaves only a small wet spot on the surface but it has thoroughly saturated a 4-foot-diameter circle; 5 gallons of rainfall spreads over a 4-foot diameter circle about 1 inch deep. So on deep, clay soil, more than 5 gallons per application may be in order.

I can't specify what is optimum in any particular situation. You must consider your own unique factors and make your own estimation. All I can do is stress again that the essence of water-wise gardening is water conservation and not leaching the root zone.

Chapter 4

Starting Seeds

Direct seeding—sowing seeds straight into the bed—is easier and much less time consuming than raising transplants. It is also far less costly than buying transplants and can be entirely reliable. Transplanting has down points; I'll list them. Transplanting a species that makes a taproot breaks its taproot and makes the plant less able to forage for moisture. By direct seeding you can know for sure which variety you're growing. This isn't always the case when you purchase transplants, which is a sad commentary on honesty in business. Buying transplants is far more costly than buying a packet of seeds. Raising your own transplants means being ready to water them daily.

Direct seeding works best when the bed's surface inch has been converted into something like potting mix by hoeing a half inch of compost into the surface inch before raking out a fine seedbed *and when the seedbed can be watered every day or two until germination occurs.* The major risks when direct seeding are sowing into soil that is too cold (or too hot) and using weak or dead seed.

The result is far less certain when germination depends on naturally occurring soil moisture.

Early Spring: The Easiest Unwatered Garden

Soil moisture loss is slow early in spring. The soil is usually damp and seeds adapted to chilly soil usually germinate without irrigation. However, some spring vegetables finish early in summer and can experience moisture stress. To reduce their need for irrigation, position them farther apart than conventional wisdom suggests these days. Spring vegetables may drain the subsoil of much moisture, so starting summer vegetables on beds that already grew an early crop may prove impossible without irrigation if late spring or early summer rainfall is scant.

Later in Spring: Sprouting Seeds Without Watering

When I performed the experiments that lead to writing this book I grew vegetables as though I had no possibility of irrigation. In Cascadia there is little chance of significant rainfall after mid-May and even less in June. I knew it would be impossible for me to make small seeds germinate after the season warmed up. To overcome this challenge I started all small-sized seeds by mid-spring.

Sprouting big seeds without irrigation—corn, bean, pea, squash, cucumber, and melon—is less challenging because these are sown deeply where soil moisture still resides after the surface has dried out. And even if it is so late in the season that the surface few inches have become dry, a narrow deep "V" made with a common hoe will expose moist soil where big seeds can be coaxed to germinate with little or no watering.

Germinating small seeds is more difficult with or without irrigation; this is the reason so many gardeners buy transplants. Spading and rototilling stops capillary moisture uplift until the soil resettles. This interruption is useful for preventing moisture loss in summer, but the same phenomenon makes a fluffy loose seedbed dry out rapidly. Successfully sprouting small seeds in warm weather is less likely without watering them.

I plan ahead. I usually prepare beds for summer crops a month ahead of sowing, thus giving the soil time to reestablish capillarity before seeds go in. The principle behind this can be easily demonstrated. Many gardeners have rototilled the soil and noticed that their compacted footprints were moist the next morning while the rest of the earth was dry and fluffy. Foot pressure restored capillarity; during the night fresh moisture replaced what had evaporated.

Here is a simple technique which helps germinate every kind of small seed. After digging or rototilling the gardener can compress the soil below the seeds and then cover the seeds with a mulch of loose, dry soil or maybe better, with sifted compost. Seeds then rest atop damp soil exactly as they lie on damp blotter paper in a germination laboratory's covered petri dish. If the seeds are covered with sifted compost instead of soil this dampness may not disappear before the sprouting seedlings have propelled a root several inches into moist soil and are putting leaves into the sunlight.

I've used several techniques to reestablish capillarity after tilling. There's an inexpensive push planter in my garage that first compacts the tilled earth with its front wheel, then cuts a furrow, drops seeds into that "V" and then its drag chain pulls loose soil over them and the rear wheel lightly packs that loose earth.

I've also used a lightly loaded garden cart or wheelbarrow to press down a wheel track and then sown seeds in it. If it is big seed I am sowing I simply sprinkle seeds in that compacted furrow and then pull loose soil over them with a common hoe. If it is small seed I carefully sprinkle a

predetermined thickness of sifted compost over the seeds. If it is small seed for what will become large plants, I place a pinch of seed in the furrow every few feet and carefully cover only that pinch with a small handful of compost that is 2 inches in diameter and ½ inch thick.

Sometimes I directly sow seeds for large brassicas and cucurbits above highly fertile positions. First, I deeply dig organic fertilizer into a circle about 18 inches across. Then with my fist I press a depression into the center of the fluffed- up mound. Sometimes my fist goes in so easily that I refill the hole with more soil and push it down again. My purpose is not to make rammed earth, but only to reestablish capillarity by having firm soil under a shallow, fist-sized depression about one inch deep. Then four to six small seeds are put in knuckle dents and covered a quarter-inch deep with fine earth or sifted compost. I often get good germination without watering. If the weather is hot and sunny such that I must water, the moisture can be confined to this spot so I only have to use a half cup of water per position. This same technique works even better on hills of squash, melon, and cucumber because these large-seeded species must be planted quite a bit deeper and rarely need to be watered. I have more to say about starting cucurbits in the next chapter.

To make finely sifted compost, I use a gold panning sieve with a 3/16 inch mesh whose diameter is the same as a 5-gallon plastic bucket. I put a shovelful of compost on the screen, hold the sieve firmly on top of a bucket and shake out fine compost.

Summer: How to Fluid Drill Seeds

Soaking seeds before sowing them is another water-wise technique. At bedtime, place the seeds in a half-pint mason jar, cover the jar with a square of plastic window screen held on with a strong rubber band. Start soaking the seeds at bedtime; drain them first thing in the morning; gently rinse the seeds with cool water; drain again and rest the jar on its side so as to allow more air exchange. Do this two or three times daily until the root tips begin to emerge—which should happen within three days. If it takes longer than four days to see root tips on most species you will have discovered in advance that you are trying to germinate dead or nearly dead seeds. It is far better to learn that before sowing and then it would be to wait two weeks after sowing them to find out. Sprouting seeds must be put in the ground as soon as root tips appear because the emerging roots become increasingly subject to breaking off as they develop and soon form tangled masses. Pre-sprouted seeds may be gently blended into crumbly moist soil or fine compost and this mixture gently sprinkled into a furrow and covered. Pre-sprouting seeds leads to a far higher germination percentage and if you choose the correct quantity of seed for the length of furrow there will be much less thinning to be done. I find that a generous tablespoonful of strong lettuce or carrot seeds will result in one seedling per inch in 50 feet of row.

If the sprouts of the species in question are particularly delicate or you want a very uniformly spaced row you may imitate what commercial vegetable growers call fluid drilling. Heat one pint of water to the boiling point. Dissolve into it 2 to 3 tablespoons of ordinary cornstarch flour.

Place the mixture in the refrigerator to cool. Soon the liquid will become a soupy gel. If it ends up too watery to suspend the seeds, heat it again and add more starch. If it is too stiff to easily mix in sprouting seeds, heat it again and add more water. Gently mix cooled starch gel with the sprouting seeds using a soup spoon. Make sure the seeds are uniformly blended. Pour the mixture into a 1 quart plastic zipper bag and, scissors in hand, go out to the garden. After a furrow has been prepared—with capillarity restored if possible—cut a small hole in one lower corner of the plastic bag. The hole size should be less than 1/4 inch in diameter. While bending over and holding the bag opening a few inches above the furrow, walk quickly down the row, dribbling a mixture of gel and seeds into the furrow. Then cover with fine moist soil or sifted moist compost. You may have to experiment a few times with cooled gel minus seeds until you divine the proper hole size, walking speed, and amount of gel needed per length of furrow. Pre-sprouted seeds come up days sooner and the seedlings will be much more uniformly spaced and easier to thin. After fluid drilling a few times you'll realize that one needs quite a bit less seed per length of row than you previously thought.

Establishing the Fall and Winter Garden

Germinating seeds in the heat of summer for fall and winter crops can be difficult. Even when the entire garden is well watered, midsummer sowings require daily attention and frequent sprinkling. However, once seeds have germinated in a regularly irrigated garden the seedlings growing usually require no more water than the rest of the garden gets. Establishing small-seeded vegetables without regular watering may seem next to impossible in hot weather. A related problem is finding enough space for both summer and late crops.

The nursery bed may solve both these problems. Instead of irrigating a large part of the garden, the seedlings start out in an irrigated nursery bed only occupying a few square yards. They can be dug up and transplanted in autumn after summer's heat ends. Were I desperately short of water I'd locate my seedling nursery against the east side of a building where it was protected from wind and got only morning sun. I might sow a week earlier to compensate for the slower growth.

Seedlings in small pots and trays require frequent attention. Fortunately, growing seedlings in little pots may not be necessary if they can be dug up and transplanted when temperatures are cool, the sun is weak and transpiration losses are minimal. My transplant nursery is laid out in rows about 8 inches apart across a raised bed. The seedlings are thinned far enough apart that when they are large enough to transplant they may be dug out without too much root system damage. When the prediction of a few days of cloudy weather encourages transplanting, the soil around the seedlings is first cut with a large sharp knife and then they are lifted, usually with a trowel. If the main growing beds are dry when transplanting must be done then a relatively small amount of irrigation can be used to moisten only the planting spots.

Where winter gardening is possible, a nursery bed helps find room for autumn/winter crops because late season transplants can be set out after hot weather vegetables such as squash, melons, cucumbers, tomatoes, potatoes, and beans are finished.

Vegetables That Must be Heavily Irrigated (may not be suitable for dry gardens.)

Celeriac	Celery
Chinese cabbage	Lettuce (summer)
Radishes (summer)	Scallions (for summer harvest)
Spinach (summer)	

Chapter 5

How to Grow It: A-Z

Some writers tell the reader what it is that they are about to tell the reader before they tell the reader whatever it is they are about to tell them. I consider that technique as much an example of poor craftsmanship as the previous sentence is. I avoid the practice. But writing rules should be broken occasionally (such as not starting a sentence with "but" or "and"), which is why it is far better to guide one's life according to ethical considerations rather than by rigid adherence to morals. In this chapter I am trying to provide insights and crucial distinctions that'll guide you to taking effective actions. A lot of what follows may not apply to you. Some of it could apply but you will miss the significance of it because you lack experience. Some of what I say in this section could be incorrect because I lack experience growing vegetables in your climate. Please do not become impatient. There may be gems buried in this compost heap that will prove well worth finding.

First, a Word About Climate Zones

The 1993 edition of *Waterwise Vegetables* was entirely focused on Cascadia where climatic differences are minor as one goes from Northern California to the Lower Mainland and Islands of British Columbia. The entire region features a rainy mild winter punctuated by occasional frosty mornings; there is very little if any rainfall during a warm summer that normally inflicts only a few very hot days. When spring planting time arrives the soil is saturated with moisture as deeply as there is soil. The entire region provides the possibility of winter gardening but the topsoil will be dry in July and early August when most winter crops are started. Because conditions are so similar throughout the region, the 1993 edition could suggest a great many excellent vegetable varieties. When the entire bioregion is viewed from the middle of Oregon, which was where I lived then, the main difference in sowing dates from Northern California to British Columbia is only a few weeks sooner or later.

Expanding this book to usefully include as much of the United States and Southern Canada as possible has been challenging. I have thought long and hard. I have maximally engaged my imagination. I have received considerable help from my publisher, David The Good, who is currently gardening in Southern Alabama, which has a climate I have visited but never gardened in. I have brooded over when-to-plant calendars from Louisiana, Tennessee and Kentucky. I consulted my own memories of growing up in Detroit, Michigan. I have had a heap of fun while writing this book! And I plead with the reader to forgive my shortcomings. To make the best use of this book the reader will have to apply local knowledge in order to correct where I am just outright wrong and to fill in important things I did not even know I did not know. In order to make my broad generalities more useful, new gardeners and those with experience who have recently moved to another climate zone should consult their state extension service about local planting dates and the names of well adapted varieties.

There are three regions in the USA and Southern Canada where winter gardening is both possible and, in my opinion, essential to providing an abundant food supply. These are the aforementioned Cascadia; the humid Deep South including Eastern Texas; Southern Arizona and those parts of California where most of its population live. I have distinguished another zone where autumn is lengthy enough to establish crops late in summer for harvest during autumn and early winter. This area includes North Carolina, Tennessee, Arkansas and possibly Eastern Oklahoma. For my own convenience I will call these states "the Middle South". And sometimes I will include the Middle South with the rest of the humid temperate eastern states.

Establishing a winter garden requires having enough soil moisture during summer's heat. In this respect the Southwest and Cascadia will require at least a little irrigation; in the Middle- and Deep South you can hope for enough rainfall later in summer. In the rest of the states an unprotected winter garden is either not possible or else extremely limited. This zone includes everything from North Carolina, Tennessee, Arkansas all the way to Southern Canada. I will name this broad area "the humid Eastern United States and Southern Canada". When gardening with little or no irrigation in this "zone" the main moisture concern involves occasionally inadequate rainfall during summer. Northern Idaho and Southern British Columbia except for the Lower Mainland and Islands might fit here too.

I am not considering the semi-arid prairies, the wheat belt and the western States including California (except perhaps where redwood trees grow). In these districts irrigation may be essential to grow much in the way of vegetables. Part 2 of this book will help gardeners in these places.

Any time I state what I do, I am describing gardening in a climate like Western Oregon, and I hope an imaginative reader can adjust that information to their own situation.

About Varieties

As recently as the 1930s, most American country folk did not have electricity. Most depended upon hand-pumped or hand-carried water. Some had wind powered pumps. Water was precious and vegetable gardens had to be grown with a minimum of irrigation. Waterwise gardening was the usual thing. In the well-watered east one could anticipate several consecutive rainless weeks every summer. So vegetable varieties were bred to grow on through dry spells and traditional American vegetable gardens were designed to help them do that.

I began gardening in the early 1970s, just when intensive raised-beds were replacing traditional long rows. The cutting-edge books I read then asserted that raising vegetables in widely separated single rows was a foolish imitation of commercial farming, that commercial vegetables were arranged that way for ease of mechanical cultivation but the home gardener could do far better than to imitate the farmer. Closely planted double-dug raised beds, weeding with fingers and the use of transplants whenever possible were alleged to be far more productive and far more efficient users of irrigation. I did all these things in my first years of gardening but experience showed me that these beliefs are mostly politically correct hyperbole.

I think this is more likely the real story: Old-fashioned American gardens had to cope with inevitable spells of rainlessness. Footpaths between long rows were not a waste of space. They provided root zone from which the crops obtained moisture. Please recall I said that much more moisture transpires from a crop's leaves than evaporates from bare soil. Looked at this way, widely separated vegetables in widely separated rows that leave bare ground showing may be considered the more efficient way to use the water nature puts there. Only after, and if, these moisture reserves are significantly depleted does the gardener have to irrigate. In the humid east this doesn't often happen. The end result is surprisingly more abundant than a gardener educated on intensivist raised-bed propaganda has been misled to imagine.

Modern vegetable varieties have been bred to be highly productive as long as their soil remains comfortably moist, which means grown with abundant irrigation. These days vegetable farmers and market gardeners establish high density populations with smaller-sized quicker-maturing varieties. Actually, the larger varieties of the past were not a great deal less productive overall, but they took longer to begin yielding.

I appreciate older vegetable varieties—sprawling, large framed, later maturing, longer yielding, vigorously rooting. However, many of these old-timers are now only grown for the picture packet trade, which is the low- and cheap end of the vegetable seed business. These familiar varieties have not seen the attentions of a professional plant breeder for decades and consequently make a high percentage of bizarre, misshapen, nonproductive plants. I urge the reader to protect themselves from failures caused by low quality seed. Buy seeds from companies that focus on reliable performance. In the short list below are businesses I am familiar with that make an honest effort to always provide productive varieties that germinate well. Not being included in this list implies no condemnation. It merely means I have no current information.

- Johnnys Selected Seeds, Maine

- Territorial Seed Company, Oregon

- Adaptive Seeds, Oregon

- West Coast Seeds, B.C.

- Stokes, Ontario

- Park, South Carolina

- Vermont Bean Seed Company, Wisconsin

Plant Spacing

Maintaining a low density plant population is the essence of dry gardening. A dry garden should be laid out in long rows. The longer the anticipated gaps between significant rains then the further apart the row centers should be—somewhere between 2½ feet and 5 feet apart. Large plants like indeterminate tomato, squash, cucumber and melon should be allocated even more elbow room. The general principle to apply when making this choice is that plants should be positioned as far apart as their aboveground parts could possibly grow under ideal conditions. You can assume their roots will be *at least* equally extensive. If you have been gardening intensively then you may never have seen how large a broccoli or tomato plant can get. If that describes your situation, I suggest you increase the area each plant is allowed for its exclusive use by 25 percent per year until the plants stop getting bigger. Because so many people these days have not been properly instructed in arithmetic I will explain what I mean by 25 percent per year. Suppose you are currently growing small cabbages on positions 18 inches by 18 inches. Each plant "owns" an 18" x 18" square, or 324 square inches. To increase that by about 25 percent, position them 20 inches by 20 inches, or 400 square inches per plant. If they fill that space, next year try increasing it to 24 inches x 24 inches, or 576 square inches for each plant.

I cannot precisely state what would be the most effective spacing for your garden. How fast does your soil dry out? Are your temperatures lower than mine and evaporation less? Or is your weather hotter? Does your soil hold more than, less than, or just as much available moisture as mine? Is it deep, open and moisture retentive? And every variety is different. I suggest you become water-wiser by testing a range of plant densities.

I also suggest that you at very least *study* all the root system drawings found in John Weaver's *Root Systems of Vegetable Crops* and meditate deeply on the implications between the lines. Some of them are reproduced in this chapter. The grid each drawing shows is 1 foot by 1 foot. Weaver's book is available for free download at soilandhealth.org.

Arugula (Rocket)

A few peppery arugula leaves in a green salad makes it much more interesting; this is especially so of cabbage salads when they get lightly dressed with olive oil and vinegar instead of being heavily coated with mayonnaise. Small young arugula leaves are mild enough that they can make a tasty salad all by themselves when dressed with fragrant virgin olive oil and shaved Parmesan cheese.

Arugula is very frost hardy. Its seeds germinate well in chilly soil and do not require super fertile soil. The seeds germinate very quickly, usually within four days of sowing.

During the long days of June and July arugula goes to seed after growing barely large enough to snip a few leaves from. Bolting to seed happens even faster when the temperatures are high. Once seed making starts I find the taste becomes intolerably harsh. In the humid eastern states arugula is a spring crop. Where autumn is extended and gentle it can be sown again a few weeks before the end of summer. In the Deep South and in Cascadia arugula can also be a winter crop. In the Middle South how long it'll stand into the winter is uncertain and depends on

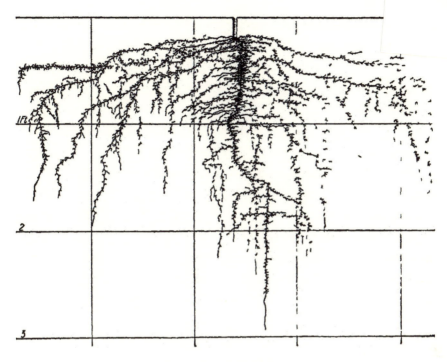

Two month old bush bean

location. Tasmanian summer days rarely exceed 80° F so I sow a few row feet of arugula every three to four weeks from early spring through early autumn. A little strip of soil may have to be watered until the seeds germinate. The species makes a strong and fast penetrating taproot so after the seedlings have formed two true leaves it may continue growing even when the topsoil is dry. I sow 8 row feet of arugula late in summer and again in autumn while the soil is still warm enough to germinate the seeds. Both these sowings usually overwinter productively. A fall sowing may not go to seed until March.

Beans of All Sorts

Heirloom climbing beans make extensive foraging root systems. Most modern bush varieties have puny root systems and probably should be avoided where summer rainfall is scant or non-existent. Perhaps older varieties like Contender or Provider are better in this respect. Bean seeds,

be they bush, pole or runner, are sown after the danger of frost has passed
and soil temperature measured an inch below the surface stays over 60°F.
If the topsoil is getting dry, at very least soak the seeds for eight hours in
tepid water before sowing them or better, chit them for a few days and
sow them when roots begin to emerge. Carefully place seeds 2 inches
apart in a deep furrow that reaches moist soil. However deep that "v"
has to be, do not cover the seeds more than 2 inches.

Thin bush varieties to one foot apart in the row. Sow climbers in
bunches of three seeds 16 inches apart and thin gradually to the best
plant in each position by the time the vines start running. Allow at least
2 feet of bare soil on either side of the climbing bean trellis for the crop's
exclusive root zone in order to avoid moisture competition from other
plants. If part of the garden is sprinkler irrigated locate the climbing bean
trellis toward the outer reach of the sprinkler's throw. Due to its height,
the trellis intercepts quite a bit of water and dumps it at the base.

You can also use the bucket-drip method and fertigate the beans,
providing around 1 gallon per row-foot every few weeks. In sand that
might mean draining one bucket per foot. In heavy ground it might be
3 gallons from buckets set 3 feet apart down the row. Pole beans will

survive moisture stress but will yield less. Any of the older pole types like Kentucky Wonder, Blue Lake Pole or especially Rattlesnake seem to do okay. Vermont Bean Seed Company uniquely offers the brown seeded version of Kentucky Wonder, often called "Old Homestead". This variant is especially drought tolerant. Runner beans seem to prefer cooler locations but are every bit as drought tolerant as ordinary snap beans.

Heirloom dry beans (sometimes called "shelling beans" when the seed is eaten before it dries down) were mostly half-tall vining types that make vigorous root systems. When grown unirrigated their seed yield is lower but the seed is still plump, tastes great, and sprouts well. The aforementioned Kentucky Wonder Brown Seeded has tasty seeds that make excellent dry beans. Black Coco is the best-tasting bush dry bean cultivar I have ever grown; it is offered by both Territorial and Vermont Bean Seed Company.

Serious self-sufficiency seekers producing their own legume seeds should consider the fava, garbanzo, lentil, cow pea, capucijner and Alaska pea. Favas can be autumn sown in regions with mild winters. In Oregon I'd sow favas when the fall rains start. They grow slowly over the winter and then grow very rapidly in spring; the seeds dry down in June. June/July in Cascadia is the best season for maturing seeds so fava beans are Cascadia's ideal self sufficiency legume. Territorial still offers Sweet Lorane. I bred this variety in the 1990s. It is a high yielding small-seeded fava that is far more cold hardy than the large seeded types. Most small seeded fava varieties are used as green manures. Their seeds taste more like animal food than human food. Not so of Sweet Lorane.

Garbanzos and lentils make small tidy bushes; they are mildly frost hardy and grow like peas. There are a great many garbanzo varieties. I was gifted with one by Carol Deppe called the Hannan Popbean. It was eaten like parched corn seeds (corn nuts). You can find it for sale online. My favorite garbanzo has black seeds and is called kale chana. It originated from Afghanistan. You can find this one at Prairie Seeds in Saskatchewan, https://prairiegardenseeds.ca/products/black-chickpea. Alaska peas and capucijners are used for pea soup. They're spring sown and grown like ordinary shelling peas. Cow peas thrive in hot humid conditions. David The Good, the publisher of this book, uses cow peas as a hot season green manure crop.

Beet

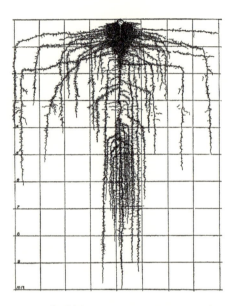

3 month old beet grown on prairie soil in Eastern Nebraska.

Where the subsoil is open and not highly acidic, beets form roots that go far deeper and wider than most people realize. In Cascadia sow them in spring after the soil temperature has reached 50°F. Beet seed germinates well in moist, cool soil. It does not germinate at all well when soil temperature in the surface inch exceeds 85°F, which can happen on a sunny day when the air temperature is below 80°F.

In the humid east above latitude 40° sow beet seed from spring until early June when germination is easy to achieve and the soil is usually naturally moist. There is little sense in starting after mid-July because by mid-August solar intensity drops so that the plants can not manufacture surplus sugar to rapidly swell their bottoms.

A single row can exhaust subsoil moisture from an area 3 feet wide. If given plenty of growing space most varieties steadily grow ever larger but will remain tender all summer. If they get so crowded that growth slows the beets will get woody and lose sweetness. That's why correct and prompt thinning/harvesting is essential. When the leaves are 3 inches tall the young plants will be securely established and there will be no more mysterious disappearances. At this point thin the row to 3 inches apart.

We usually eat these tender leaves in salads; they may also be steamed and eaten like delicate Swiss chard. When they've made golfball-sized bottoms, thin the row to 6 inches apart and eat baby beets. When they reach the size of large lemons, thin to 1 foot apart. Given this much room and deep, open soil, many varieties achieve the size of soccer balls by the end of summer.

In Cascadia fertigation should not be necessary when they grow in deep, open soil. I've done especially well dry gardening Early Wonder Tall Top, which is a true heirloom; when EWTT is large, it develops a thick, protective skin and retains excellent eating quality. These days I use a "baby" variety. This does not mean it only makes small beets. It means small plants quickly make edible parts the size of golf balls which get ever larger as the plant grows.

In the Deep South sow beets in February and again in September. If you start later you won't get much to eat. In the Middle South sow them in March and again in the first half of September. Make the late summer sowing during a period of cooler weather. Throughout the South, autumn-sown beets will not achieve such large size so they need not be thinned so severely.

Broccoli: Italian Style

Italian broccoli needs abundant soil moisture to make tender large flowers. Given enough elbow room, many varieties can endure long periods of moisture stress, but under moisture stress the smaller, woody, slow-developing florets won't be great eating.

In Cascadia and in the humid eastern half of the USA and Canada, directly seed broccoli in spring as soon as reliable germination can be obtained. In Cascadia sow a new patch every six weeks until mid-July. Plan to fertigate. Do the same in the humid eastern states although fertigation may not be necessary. In the Middle- and Deep South broccoli is a cool season crop, grown in spring and started again towards the end of summer. In the mildest parts of Southern Cascadia Italian broccoli varieties may be started toward the end of summer. They will make flowers during winter.

Broccoli tastes best when big plants can form flowers that are way too large for the supermarket trade. I routinely harvest broccoli the size of dinner plates. To achieve that I allow the plants a 4-foot-wide row and position them about 3 feet apart in the row.

I know how to raise stronger brassica transplants than I can buy at the garden center—but I never transplant broccoli. Never! Transplanting almost inevitably checks growth for a week while directly seeded plants form a much stronger root system. Also, a lot of effort and time can be saved by sowing seed directly compared to what it takes to raise transplants. Also, I am not accusing any particular seedling nursery business of deceptive or fraudulent practices but by starting from seed I know for sure which variety I am growing. My favorite variety for spring sowing is Amadeus from Johnnys. I think the best variety for late summer and autumn harvest is Arcadia, also from Johnnys. Both these varieties make big very delicious main flowers followed by lots and lots of large side shoots. Do not choose compact varieties; they will have stunted root systems.

I start all the big brassicas the same way—broccoli, cabbage, Brussels sprout, kale and cauliflower. For each plant I make a low mound of super soil about 16 inches in diameter. First I spread one shovelful of compost and some fertilizer over a circle 12 inches in diameter and dig it in. The spot ends up loose and fluffy, like potting mix. Then I gently press it down in the center with my fist to restore capillarity and put 5 seeds there. If your soil is clay no amount of compost will turn it into potting mix so consider also buying a bag of prefertilized seedling raising medium from a garden center. After you have made a super fertile spot in that clay, make a small hole the size of a half pint mason jar in the center of each position. Fill that small hole with seedling mix and pat it down firmly to restore capillarity. Then sow five seeds in the seedling mix. Bagged prefertilized seedling mixes from the garden center often are not all that fertile. If early growth is not rapid, fertigate these positions.

Thin each position gradually to the best single plant; complete thinning by the time four or five true leaves have developed. When the topsoil gets dry fertigation every two weeks makes an enormous difference. You may be surprised at the large size of the heads and the quality of the side shoots.

Many broccoli varieties have weak root systems. I'd avoid any variety that makes medium-sized heads held up on a tall stalk for convenient mechanical harvesting, or one that is compact, suitable for growing in containers on a patio. Neither of these types will make many side shoots. Look for varieties said to be late maturing (forms bigger plants); these make large flowers and usually form more—and larger—side shoots. If spring sowing then find varieties said to handle summer's heat.

Grow hybrids! Open-pollinated varieties like Italian Sprouting Calabrese, DeCicco and Waltham 29 have catastrophically degenerated. They once were widely used commercial varieties before hybrids took over the market. They now are highly variable, make coarsely beaded flowers with second-rate flavor and throw small rather woody side shoots. Dry gardeners who want side shoots for as long as possible during summer may prefer crude, open-pollinated varieties.

Broccoli: Purple Sprouting

This type of broccoli (there are many variants from England) is considerably more cold hardy than the Italian varieties usually found in supermarkets. I believe purple sprouting originated as an Italian x kale cross. The flavor of purple sprouting is not as gentle as Italian broccoli and the flowers are smaller (and more numerous).

Purple sprouting usually survives Cascadian winter when Italian sorts usually do not. In my opinion purple sprouting types do not taste quite as good as Italian broccoli but in Cascadia any fresh garden food during the months of March and April is a treasure. Purple sprouting is strongly biennial. No matter if it is sown in spring, summer or autumn, the plants must first overwinter before blooming. This vegetable will not survive a freezing winter in the humid east. It has survived nighttime low temperatures below 10° F in Oregon so I speculate it might live through winter in the Middle South if the soil does not freeze solid.

When dry gardening purple sprouting types in Cascadia it is easiest to establish the plants in spring. Given fertigation and almost a year to grow before blooming Purple Sprouting may reach 4 to 5 feet in height and 3 to 4 feet in diameter, and yield hugely. It is not essential to heavily fertigate Purple Sprouting through the summer although with generous fertigation you can truly grow enormous plants for their beauty. Quality or quantity of spring harvest won't drop one bit if the plants become a little stunted and gnarly in summer, as long as you fertilize at the end of summer to spur rapid growth during fall and winter.

I am sure purple sprouting varieties will perform in the Deep South when sown at the end of summer. But there is no reason to grow them in the Deep South because Italian varieties taste better.

Root System Vigor in the Cabbage Family
Wild cabbage is a weed and grows like one. In a pasture it is able to successfully compete for water against grasses and other herbs. Just for fun, I once grew a wild cabbage weed, hoeing the area around it free of weeds, generously fertilizing it in spring but giving the plant no irrigation; it ended up 5 feet tall and 6 feet in diameter

As this highly moldable family is inbred and shaped into more and more exaggerated forms it loses the ability to forage. Kale retains the most wild aggressiveness, Chinese cabbage perhaps the least. Here, in approximately correct order, is shown the declining root vigor and general adaptation to moisture stress of cabbage family vegetables. The table shows the most vigorous at the top, declining as it goes down.

Adapted to dry gardening	Not vigorous enough
Kale	Italian broccoli (some varieties)
Brussels sprouts (late types)	Cabbage (small supermarket types)
Late savoy cabbage	Brussels sprouts (early types)
Giant "field-type" kohlrabi	Small "market-garden" kohlrabi

Adapted to dry gardening	Not vigorous enough
Mid-season cabbage	Cauliflower (regular, annual)
Rutabaga	Chinese cabbage
Italian Broccoli (some varieties)	

Brussels Sprout

In case you have never grown Brussels sprout, I'll mention here that the plant resembles the tall species of kale, which is, I believe, what it was bred from several hundred years ago. It forms a tall strong main stem with opposing pairs of large leaves every few inches up the stalk. A well-fertilized plant given adequate moisture and growing room can be waist-high at summer's end before it starts forming sprouts.

Like kale, Brussels sprout develops a vigorous wide-spreading root system. They need less fertigation than most big brassicas require. Planting in spring leads to sprouts forming well before the end of summer; this approach does not work because the sprouts themselves must develop in cool weather or else they will be oversized, not internally tight, will taste harsh and are often ruined by aphids. The sprouts fatten up over several months, starting at the bottom of the stalk and proceeding to the top. If the early-forming sprouts are ruined by aphids and/or too much heat (and removed from the plant) then higher-quality sprouts may still form farther up the stalk during autumn (and during winter in districts where winter gardening is possible). But why waste a third or more of the potential harvest!

The challenge is to find the correct starting date for your location that makes sprouts start fattening in early autumn. In Cascadia it is best to sow a late maturing variety around June 15. That way the first sprouts develop in mid-October, hopefully after there have been a few frosts which eliminates almost all the aphids. Cascadians will have to spot water the positions to germinate the seeds and then fertigate as often as necessary to keep the plants growing fast. Gardeners in the humid eastern states will discover their best planting date may be a few weeks earlier or later than mid-June.

In general, grow Brussels sprouts like Italian broccoli. If you anticipate moisture stress then position plants 3 feet to 4 feet apart in the row. If fertigation is provided every few weeks during July and August the plants may be 3 feet in diameter and 4 feet tall by October and yield enormously.

Except in states bordering Canada and in Southern Canada gardeners in the humid eastern states should choose late maturing European hybrids like Johnny's Diablo or Divino, Territorial's Nautic or West Coast's Igor. Closer to the Canadian border autumn is brief and the harsh winter soon kills the plants. There try earlier maturing varieties that form sprouts over a shorter period.

In the Middle South Brussels sprouts are started in August. In the Deep South start them early in September. Use early varieties, like Jade Cross or Oliver.

Cabbage

Unless you are confident there will be plenty of soil moisture during the entire time they are growing I suggest you forget about small (quicker maturing) supermarket-sized cabbages—red or green. Varieties that need more growing weeks in order to produce larger heads will do much better in the humid east, especially if they are given minimal fertigation during dry spells. Seven years of growing commercial variety trials for Territorial Seed Company taught me that savoy varieties are the tastiest and most resilient.

In the humid eastern half of the United States and in Cascadia I suggest delaying the first sowing until mid-spring. These will be harvested during summer. I suggest using mid-season varieties for this; the traditional one is called Copenhagen Market. Johnny's sells excellent late cabbage varieties such as Deadon and Clarissa and suggests sowing them in May and June, how late depending on how long they'll survive autumn.

Except in the hottest parts of Southern Cascadia sow again when summer first arrives; if I have to pick a date for this I'd say June 15th and urge the gardener to choose late and very late varieties bought from

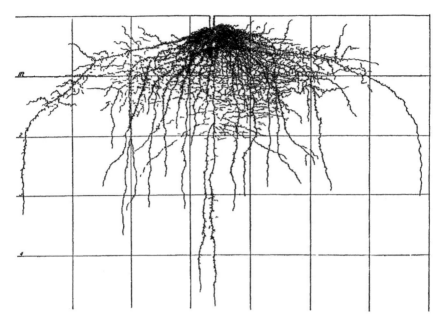

Mid-sized cabbage at 75 days

Territorial or West Coast. These varieties tend to be much later than the later varieties sold by eastern seed companies like Johnnys. This later sowing is for harvest late summer, fall and hopefully well into winter.

In Southern Oregon and Northern California delay sowing cabbages for autumn/winter harvest until mid-July. Spot water the positions when germinating seeds and fertigate as needed.

Where winter gardening is possible a unique type of cabbage can be sown at summer's end, it overwinters and makes heads in spring without going to seed first. The English call this type "spring cabbage" and grow it in their mildest districts located near the English Channel. In Cascadia a nursery bed of this sort may be sown in August (or in September in Northern California), dug up with a trowel and transplanted out during autumn. Transplanting is very easy and secure in this season. British market gardeners have access to many spring cabbage varieties. They consider overwintered cabbage to be a risky crop because the local varieties often make seed stalks instead of useful heads. My favorite variety of this type was bred in Japan and is called Spring Hero. It is entirely dependable. Spring Hero absolutely does not go to seed before forming heads. In fact, most of the plants will not make seed at all, ever. In greater fact, if the first head that forms early in spring is carefully cut off such that the basal leaves remain on the stem, then the plant will reshoot a ring of what will become little cabbages around the stump. If all but one of them is removed after they have grown to about an inch across, the one survivor will form another large head in about six weeks. And if that process is repeated yet again a third head may form late in spring. However, harvesting must stop after the third head because long days cause the cabbages to form distorted shapes, not properly fill out and lose flavor. I suppose I should also mention that Spring Hero is the best tasting, most tender salad variety I have ever grown. I used to sell Spring Hero seed when I owned Territorial Seed Company but the current owners have no interest in spring cabbage. I jump Australian quarantine hurdles in order to buy seeds for Spring Hero from Molesseeds.co.uk. Moles is a wholesale business only serving farmers and market gardeners; fortunately their smallest size packet provides the largest amount I can use up during the years the seed will remain viable.

In the Deep South cabbage is a cool season crop established from early August through January. In the Middle South cabbages for autumn harvest can be started when summer's heat declines. Tennessee Extension Service says 1 September is the latest sowing date but if you're growing larger varieties that need more growing days I'd start no later than mid-August. Because of the reduced rate of soil moisture loss during that season I expect that small supermarket cabbages will grow well on rainfall although the seeds might need a bit of spot watering to help them germinate. In the Deep South Spring Hero might deliver a succession of heads if sown in September.

Quick-maturing varieties make small plants; these may be separated by about 30 inches. Later maturing varieties make larger plants. If moisture stress is anticipated these should be given 3 feet of in-the-row space, 12 square feet of growing room per plant. Sow and grow them like broccoli.

The more fertigation you can supply, the larger and more luxuriant the plants and the bigger the heads. Gauged by increased yield from water expended, it seems well worth it to fertigate late varieties.

Japanese savoy hybrids like Savoy King make especially tender salads. Stokes Savoy Ace is the only Japanese savoy I can conveniently find offered in North America. European savoys are readily available through US garden seed companies but are coarser, thicker-leaved, and tougher than Japanese varieties.

Carrot

During its first month a carrot seedling puts more energy into growing a tap root than it does making new leaves. If the tap root is not terminated by a clod, stone, a hot bit of fresh chicken manure or an undissolved chemical fertilizer granule and if commercial quality seed was used then the part we eat will also be straight and full length. Before sowing I usually spade the carrot bed 10 inches deep; if I am very busy I rototill, but my old front-end tiller only loosens about 8 inches deep, at best. Carrot seeds germinate poorly in cold soil; the seedlings grow very slowly in chilly weather and this slow start inevitably lessens quality at harvest. Mid-spring is the sensible time to establish the bed.

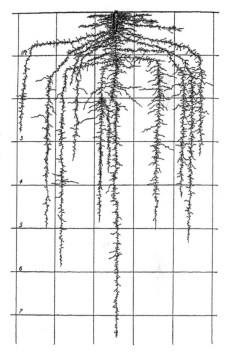

Mature carrot. Imagine what this drawing would have looked like if there had been one such carrot every foot in the row, with rows 4 feet apart.

When I grow carrots with plenty of irrigation I arrange them in short rows across a four-foot-wide raised bed and make the rows 18 inches apart. I start five or six such rows every month starting mid-spring and continuing through mid-summer. The last sowing involves three times as much space because this one has to feed us through winter. Carrot seeds are slow to germinate, they are small and must be positioned no more than ¼ inch below the rapidly drying out surface. I hand water the carrot bed until the seeds sprout. In sunny warm weather I do that daily. I promptly thin the rows so the carrots stand ½ inch apart before they are ¼ inch in diameter. A few weeks later I thin out every other carrot in every row; we eat these as "baby" carrots. I thin the whole patch at one go. The survivors now stand at least an inch apart. Then I harvest alternate rows so the between-row space becomes 36 inches. I harvest this way not to manage soil moisture but because it allows in more light and because large carrots often must be dug and cannot be pulled out by their tops. This approach works great when the soil is always comfortably moist.

In Cascadia carrots can be grown *entirely* without irrigation but you only get one shot at it in mid-spring. This one patch has to supply the household through summer, autumn and the entire winter. It works because carrots do not necessarily become woody and tasteless when they get quite large—unless crowding makes them stop growing. Allocate 4 feet unoccupied width to a single row. When the seedlings are about 2 inches tall, thin carefully so the plants stand 1 inch apart. When the carrots are ¼ inch to ⅜ inch in diameter thin/harvest to 6 inches apart. Make the final thinning when the carrots are ¾ inch in diameter; at this point they should stand 1 foot apart in the row. Irrigation should not be necessary. Foliar feeding every few weeks will make much larger and tastier roots.

Choosing the right variety is very important when one sowing has to provide for the entire year. Tender munching varieties usually lack enough fiber to get large without splitting. Be prepared to experiment with variety. Look for "autumn" or "winter" varieties; these are bred to stand in a garden for months to be harvested as needed. Huge carrots are excellent in soups and we cheerfully grate them into salads. Something about accumulating sunshine all summer makes the roots incredibly sweet. In my Oregon experiments I had spring-sown carrots remain in good eating condition until next March.

Sadly, I have to point out here that because of carrot rust fly larvae gardeners around Puget Sound will likely not succeed with carrots from mid summer through winter unless they construct fly screened cages or maintain a fly-tight floating row cover over their carrot bed.

Dry gardeners in the humid east will do something between the two extremes I have just described. I suggest using a pair of rows that are 1 foot apart with 4 feet from their center to the next row center. Sow mid-spring. Thin these carrots 1 to 3 inches apart, how far apart depends on the likelihood of moisture stress and how long you want them to be able to keep getting larger. If the topsoil is moist in mid-June, it will be possible to start another patch for autumn/winter storage. The pre-sprouting techniques described in the previous chapter will help these slow-to-germinate seeds to emerge far sooner.

In the Deep South carrots may be sown in September/October and again late January through mid-February. My reading of monthly rainfall charts in the Deep South suggests this crop will not likely experience moisture stress. I suggest thinning 1- to 2 inches apart in rows across a 4-foot-wide raised bed. Make the rows 18 inches apart.

Cauliflower

Cauliflower makes a compact rather shallow root system with a limited ability to forage. Worse than that from a dry gardener's perspective, even moderate moisture stress at any time during the growth cycle, even when the plant is very small, prevents the plant from heading properly. Without irrigation, spring-sown cauliflower should be started early and be a variety that matures quickly so they're harvested before intense sunlight and heat cause a high demand for moisture. Besides, cauliflowers do not taste great when they mature in hot weather. For that reason alone I do not schedule cauliflowers for summer harvest. For autumn harvest in

Cascadia use late maturing varieties. At Elkton, Oregon I started them mid-June through mid-July. These needed fertigation. In the Middle and Deep South cauliflower may be started late in summer and again early in spring. I do not know if gardeners in the humid eastern states should use quick maturing or late varieties for harvest at the end of summer. I suppose it depends upon how quickly winter arrives and how harsh autumn will be.

The very best varieties for dry gardening in Cascadia are overwintered types that are sown during the difficult heat of early August and form heads the next spring. These must be spot watered if they are to grow large enough to survive winter. Delaying sowing until September or early October when the plants will have less need for moisture only works in Northern California where winter is very much milder and more often

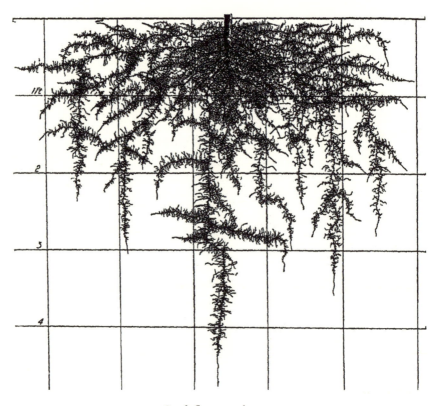

Cauliflower at harvest

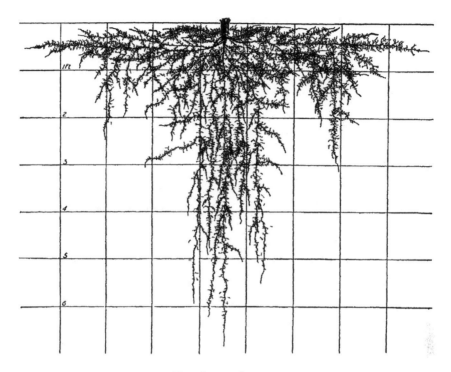

Chard at midsummer

sunny. When water is very short try sowing overwintering varieties in an irrigated nursery bed and transplanting out seedlings in October. Dig them carefully from the nursery bed with a trowel and position the seedlings either 2 feet apart in rows 3 to 4 feet apart or on wide raised beds spaced 24" by 30". There should be no worries about moisture stress before they form heads. The only overwintering varieties available in North America are sold by Territorial Seed Company and West Coast Seeds who import the seed from Holland and England.

Chard

This vegetable is a type of beet that makes succulent leaves instead of edible roots. It is far more drought tolerant than beetroot but if you want heavy leaf production during a dry summer, you may have to fertigate them occasionally.

A look at the root system drawing should show you why it makes sense to position individual plants at least 24 inches apart in the row. When grown entirely without irrigation in Cascadia I allowed each plant to exploit a space 4 feet by 4 feet. In Cascadia the spring sowing can be harvested through summer, autumn and the plants usually survive winter before they make seed the next spring.

Like beetroot, chard seeds fail to germinate in hot soil. For autumn/winter harvest gardeners in the Deep- and Middle South can start chard after the heat of summer has passed. This crop should need no irrigation except perhaps to sprout the seed. In Cascadia plants started this late in summer do not achieve enough size to yield much in autumn/winter.

The red chard varieties I have observed in my own variety trials are not suitable for starting early in the season; they have a strong tendency to bolt prematurely when spring sown.

Cilantro/Coriander

Everything said above about arugula applies to cilantro, except that its seeds need slightly warmer soil to germinate than arugula does. These days some garden seed companies offer unique varieties. I always save money by sowing seeds from the bulk herb/spice section of the health food shop because in my experience, when the seed seller's hype says a cilantro variety is slower to bolt to seed, it only means a few days later.

Corn

Americans seeking food self-sufficiency should take inspiration from First Nations people. In what had not yet become the Eastern United States their basic staple food was grain corn grown in large gardens. They improved the soil by mulching with leaves gathered from the surrounding forest floor. And every (I hope this is still true) American schoolkid learns how Squanto taught the Pilgrims how to grow corn by burying a fish below a little mound (hill) and planting four seeds over it—one for the worm; one for the crow; one to rot; and one to grow.

The object was to end up with one corn plant in each position; the positions were on a 4 foot by 4 foot grid. Native American corn varieties formed multiple stalks; thus each position made several ears. Spaces between corn plants were covered with squash vines, both winter and summer types. Runner beans twined up corn stalks (they mainly ate the dry seeds, not the pods). Harvested corn is easy to protect from rats and mice because the cobs can be hung in braids that are formed by peeling back and plaiting the attached wrapper leaves.

These days we do it differently. I suggest spreading fertilizer/compost and rototilling (or digging) the corn patch as early in spring as the earth can be worked without making many clods. This lets spring rains transport plant nutrients deeper, where the soil will remain moister for longer. For corn that is more important than making a fine seedbed; the big seed germinates well even in rather rough ground. In the humid eastern states and in Cascadia sow the seeds after frost danger ends. Farmers these days are advised to delay sowing corn until the soil temperature has warmed over 50° F. But going for an early start is risky because a period of chilly wet weather can cause poor emergence. You will see a much stronger start if you wait for warmer soil. I suggest delaying until the soil temperature (measured by a soil thermometer two inches below the surface) is ideally at 60° F—65° F. If the soil was prepared well before sowing you can afford to wait for it to get properly warm because there will be few weeds exhausting soil moisture. If many weeds do come up before sowing corn seed, hoe them out promptly or rotary cultivate shallowly immediately before sowing. Being large, corn seeds can be planted up to two inches deep where adequate soil moisture still exists after conditions have warmed up. If the soil seems dryish, gardeners can soak the seeds overnight in tepid water and sow the next morning. Farmers can not do that.

How long you can afford to delay sowing also depends on what your purpose is. If you are growing sweet corn there will be plenty of time to harvest before summer wanes. However, if your intention is to produce

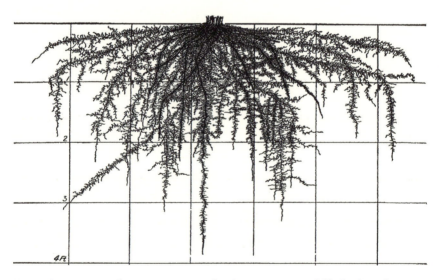

Sweet Corn at 8 weeks. Can you see why the rows can usefully be four feet apart center to center?

grain or seed corn, then in order to dry down fully in the field the crop may need nearly as much growing time as there can possibly be.

The hotter and drier summer usually is, the farther apart the plants should be. Grown entirely without irrigation on deep moisture-retentive soil at Elkton, Oregon—where it never rained in summer—I had fine results by spacing individual corn plants 3 feet apart in rows 4 feet apart, or 12 square feet per each plant. Were I gardening without irrigation where summer is rainless but not hot, say around Puget Sound and the Lower Mainland of British Columbia. I'd provide 6 square feet per plant, or 2 feet apart in rows 3 feet apart. Where I could expect summertime rainfall, I'd more or less imitate what is done in the corn belt states— plants 1 foot apart in rows 30 to 36 inches apart. Gary Nabahan describes Papago gardeners in Arizona positioning traditional varieties 10 feet by 10 feet. Lowering plant population density does not reduce yield as much as you might think. Light levels are higher at the base of the plant when corn is grown on wide spacing; this provokes tillering (which is where the plant makes multiple stalks; each stalk can make one ear and sometimes two). Most modern hybrids have been selected to *not* tiller. For that reason dry gardeners might do better with open pollinated varieties. I'd buy OP corn seed from Johnny's or Adaptive.

Cucumbers (And Additional Notes on Other Curcurbits)

Cucurbit seedlings of all types—cucumber, squash and melon—lose many roots when they are transplanted. The taproot that most cucurbit varieties form gets broken during transplanting. This makes the plant far more dependent on irrigation. Direct seeding is cheaper and usually far more successful. Germination of all species of cucurbit seeds depends on high-enough soil temperature and not too much moisture. Squash are the most tolerant to chill and moisture, melons the least tolerant. So I suggest this sowing strategy. Sow squash before cucumber. If they don't come up in a week, plant more, then keep doing this week to week until some come up. Once the first squash seedlings appear, it is time to sow cucumber seeds, starting a new batch each week until one of them emerges. When the cucumbers first germinate, it's time to try melons.

Cucumbers, squash, and melon seeds are traditionally sown into a deeply dug, fertilized spot that usually looks like a little mound after it has been dug. It is commonly called a "hill". I described how I prepare hills in the section on broccoli a few pages back. If the soil is dryish when the hill is prepared, then thoroughly water that spot immediately after digging and wait two days before sowing seeds. If at all possible do not water while the seeds are sprouting and hope it does not rain until seedlings emerge. You want the soil damp but not soaking wet. Do not water germinating seeds every day. I suggest every three days. This works because the seeds are positioned over one inch below the surface where most kinds of soil do not dry out after one hot sunny day.

Approaching cucurbits this way ensures that you'll get the earliest possible germination while being protected against the possibility that cold, damp weather will prevent germination or permanently spoil the growth prospects of the earlier seedlings.

Chitting or pre-sprouting cucurbit seed greatly speeds emergence and increases the probability of success. This technique really shines when unsettled weather has slowed soil warm-up. Cucurbit seeds must not be soaked in water. Instead, fold two separate squares of kitchen paper

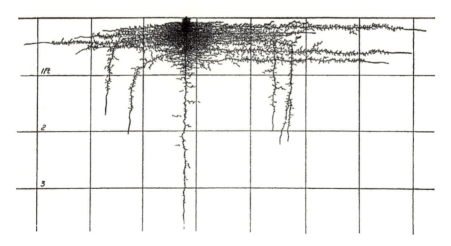

Cucumber at midsummer

toweling into quarters. Sandwich a few cucurbit seeds between them. Position each seed carefully so that the ends from which the root emerges all point in the same direction. Then moisten the paper until some water drips out and then, holding the sheets between your palms, press them together firmly. Squeeze out as much water as will easily release; don't try make the towels too dry because even ten big seeds in the sandwich absorb a lot of water. Then put this damp seed sandwich inside a small nearly airtight plastic container such as fast food comes in, or use a zipper bag. Put that in a warm place, ideally at 75°F/24°C. You can stack several seed sandwiches in the same container and keep track of varieties with small plastic labels marked with a waterproof felt pen. Position the container so all the emerging roots will point straight down. They will follow gravity. If the root does not point down when it first emerges it will form a loop or a sharp bend; that will not be helpful when it is time to put that seed in the earth. Start checking for roots after 36 hours and every 8 hours thereafter.

Sow the seeds before the root gets longer than the seed itself (which will be only a matter of hours after it first appears) because the brittle roots easily snap when being handled and this kills the seed. If there is need to delay sowing you can put the seeds into the least cold part of the fridge overnight. At each position gently place three already sprouting seeds into loose damp soil with the emerging root pointing straight down, the seed's top about 1 inch under the surface. If the soil is dry you only need to moisten an area 4 inches in diameter. You want damp soil, not soggy wet soil. *Do not waterlog the spot; do not water the spot after the seeds have been placed until leaves emerge.* Leaves will appear in a few days; these seedlings will initially grow faster than unchitted seedlings that struggled for many days before they found light.

Regular fertigation increases the yield several hundred percent and may also improve flavor. Cucurbit root systems mainly develop in the topsoil and resemble the vine. The roots will reach every bit as far as the vines grow if soil conditions allow. This is also true of most varieties of bush zucchini, which above ground forms an ultra-compact vine resembling a bush but whose root systems may spread like winter squash.

After the vine has been running for a few weeks a single fertigation position will not moisten enough of the root system. I suggest feeding two buckets of fertigation to a vine that is 4 to 5 feet in diameter, positioning one bucket on each side of the plant; feed three or four buckets to a vine that spreads out 6 to 7 feet in diameter.

I've had very good results dry-gardening middle eastern cucumber varieties, also called beit alpha or Lebanese. These make pickler-size, thin-skinned cukes that need no peeling and have terrific flavor. The burpless or Japanese sorts don't seem to adapt well to drought. Most slicers dry-garden excellently. Apple or Lemon are similar heirlooms that make very extensive vines with aggressive roots; they should be given more elbow room than most. I'd avoid any variety touted as being for pot or patio, for being compact, or short-vined, because both its vine and root system will be dwarfed.

Cucumbers do very well on trellises, though they can be left to run on the ground.

Eggplant

In my experience eggplant plants grow larger and yield sooner and more abundantly when they are grown without regular sprinkler irrigation. Perhaps this is because I grow them in a cool climate or perhaps this naturally drought-resistant tropical species does not like having its soil temperature lowered by frequent watering. Set out transplants two weeks after the tomatoes go out, well after all frost danger has passed. Make a deeply dug fertile hill for each transplant. When I dry gardened eggplant in Cascadia I positioned plants 3 feet apart in rows 4 feet apart. In Cascadia eggplant without any watering will grow a small bush that only produces a few fruit, but a few gallons of fertigation every few weeks may result in the most luxurious, hugest, and heaviest-bearing eggplants you've ever grown.

Endive (a.k.a. Escarole)

We consider endive to be a valuable salad green during autumn and into winter. How long it provides after summer ends depends upon how wintery your winter is. Much like its wild relative chicory, endive is able to grow on through prolonged summer drought. In hot weather endive tastes too bitter for me to eat so it doesn't matter if the plant struggles with moderate moisture stress through summer just so long as rapid leaf production happens in autumn. The main obstacle when dry-gardening endive is that if it is sown in mid-spring while daylengths are increasing, which is when germination of this shallow-sown small seed usually is a snap, then the plant will switch from vegetative growth to seed making before autumn. The crucial sowing date seems to be about June 1. In the humid eastern states and in Cascadia, sowings made before June bolt in July/August; but when endive is sown after June 1, maybe better just at the summer solstice, seed making doesn't begin until the following spring—if the plant survives winter.

In districts where autumn vegetables have a good chance to survive well into November it might be better to start endive early in July if there is enough soil moisture to do that. Endive seeds are small and must be planted close to the surface. Summertime germination may not happen without watering the seedbed. One solution is soaking the seeds overnight, then rinsing and draining them twice a day until they begin to sprout, and then fluid drilling them.

With irrigation in Cascadia I sowed endive in June. The plants formed huge rosettes (like oversized lettuce) by autumn. Doing variety trials at Lorane, Oregon, I recorded escarole surviving an overnight low of 6°F without any damage (in a crude tunnel cloche where it probably was a few degrees warmer). But unless the plant is protected from Cascadia's endless winter rains leaf diseases gradually rot its leaves back to the stump. I believe this susceptibility to winter leaf diseases is as much due to Cascadia's lack of sunlight as it is to high humidity and chilly conditions. In Cascadia sow clusters of 4 to 6 seeds on a spot the size of a 50 cent piece, the positions spaced 18 inches apart in rows 3 feet to 4 feet apart. You may have to spot water the positions until there is germination. Thin each position gradually so only one plant remains by the time it is about three inches in diameter. Without a drop of added moisture the plants, even as tiny seedlings, grow steadily as long as no other crop is invading their root zone. The only time I had trouble was when the endive row was too close to an aggressive yellow crookneck squash. By about August the squash roots began occupying the endive's territory and the endive was stunted. A light side-dressing of complete organic fertilizer or possibly even of compost in September combined with rain early in October will grow huge plants before winter sets in. Curly types (escarole) seem more tolerant to Cascadian winter than broad-leaf Batavian varieties.

In the humid eastern states and in the South unirrigated endive can be positioned closer together; I suggest 18 inches by 18 inches apart on wide raised beds. Where the garden is generally laid out in long rows that are 4 feet apart, make a double row with 16 inches between them and thin to 16 inches in the row.

In the Deep South endive/escarole is best sown after the summer's heat breaks. This usually occurs mid-September. Harvest should begin in about two months and continues until the plants start going to seed in spring. Endive can be sown in August in the Middle South, where they may survive winter, especially so if put under a cloche.

Herbs

Many perennial and biennial herbs like oregano and sage are actually weeds or wild perennials that thrive in semi-arid regions. In Cascadia where it almost never rains enough to matter in summer, merely giving herbs a little more elbow room than usually is offered, along with thorough weeding and side-dressing with a little compost in fall is enough coddling. Dill and cilantro are drought tolerant annuals. Basil, however, needs rich soil and considerable moisture.

Kale and Collards

Since age 30 I have lived in climates that allow winter gardens and made the maJoryty of my food be whatever the garden can produce in the season it will produce it. Living this way gradually refined my eating habits. My previous wife Dr. Isabelle Moser (died 1996) educated me about preserving health by practicing dietary self-disipline and about healing disease by dietary restriction. Thanks to Isabelle I came to like cabbage salads nearly as much as lettuce. Since lettuce freezes out during most Cascadian winters (the best winter varieties are hardy down to 18° F), learning to love raw cabbage proved very useful because savoy cabbages almost always survived winter. Kale is even hardier; it once survived an unusually harsh Cascadian winter while our savoy cabbage froze out. Kale is easier to grow than cabbage and produces abundantly in soil that is not fertile enough for cabbage while demanding less soil moisture. Twenty percent finely cut kale in a cabbage salad makes it much more interesting. So does a small amount of arugula.

You may be surprised as I was when I learned that kale produces a larger amount of complete protein per area occupied per time involved than any legume crop, including alfalfa. When roughly equal weights of kale and potatoes (weight, not volume) are steamed in the same pot, then mashed together and given some butter, the two vegetables nutritionally complement and flavor each other. The Scottish name for this delicious dish is colcannon. Food self-sufficient Cascadians could be far healthier subsisting on potatoes and kale compared to depending on the low protein soft wheat the region produces. With colcannon as their staff of life instead of bread they would need far less land to feed the family. I cook colcannon by covering the bottom of a soup kettle with potatoes cut in quarters, adding a quarter inch of water, then nearly filling the pot with de-stemmed kale leaves, steaming both until the spuds are mashable, the kale well cooked and most of the water evaporated. Then I mash both at the same time, adding salt, pepper and plenty of butter.

The key to enjoying kale as a salad component is varietal choice—and eating the tender half-grown leaves formed during cool weather. I hope all that hyperbole has interested you in trying kale. But I got to warn you—supermarket kale is usually awful! Kale needs to be consumed soon after harvest. The leaves get quite tough and the flavor becomes bitter just a few days after harvesting, even if the kale is organically grown. And when grown with chemical NPK on soils lacking organic matter, it doesn't taste great even when very fresh.

Kale also doesn't taste all that great in hot weather. In the humid eastern states fast-growing kale is best started in midsummer for harvest from early fall until it freezes out. This brings to mind one reason cabbage might be more useful than kale where winter is wintery, because cabbage can be stored in a root cellar or as sauerkraut. Come to think of it, I used to shop at a Korean market where the family sold a dozen types of homemade kimchi, including one I especially liked made entirely of young radishes, tops and all. I bet quite tasty kimchi could be made with kale.

Kale is absolutely biennial—if it is started in spring it will not bolt until the next spring. If a kale variety you sowed in April/May does go to seed the first year perhaps you should consider using a different seed company. A Cascadian dry gardener can conveniently sow kale in mid-spring. Starting early also allows a deep root system to develop before the soil dries much. There are two distinct kale species. *Brassica oleracea* types sown early grow a taller central stalk which will prove useful after the plant overwinters (more on this soon), while early sown Siberian (*B. napa*) varieties tend to form multiple rosettes by autumn, also useful at harvest time.

When I grew kale entirely without irrigation in Cascadia, I positioned plants 4 feet apart, like broccoli. The somewhat stunted plants survived summer. Plants given occasional fertigation grew lushly and nearly bumped in the row by September. If fast growing kale is started by mid-summer in the humid eastern states there will be plenty during autumn.

In the Deep- and Middle South start kale when summer's heat breaks. In the Deep South it can be harvested until it goes to seed in spring. Winter survival is less certain in the Middle South.

Attractive looking bundles of frilly large leaves are sold in supermarkets but these are the worst-tasting part of the plant. If big leaves are chopped finely enough they can be eaten raw by people with good teeth who have been convinced by health book propaganda that bitter is better. However, half-formed young leaves are more tender and mild-flavored. But only take half of them. Let the other half stay on the plant and grow old. That way you'll end up harvesting a lot more.

When the type that makes a tall central stalk (*B. oleracea*) survives winter and starts growing fast next spring every spot along that stalk where a leaf had been will put forth a cluster of tender little leaves (resembling Brussels sprouts). These are salad quality greens.

Siberian kale (*B. napa*) is low-growing, forms multiple rosettes at its base as summer progresses. The more rosettes that develop the more little leaves there will be to harvest, so there is additional advantage to starting it mid-spring when germination of the small seed is easier. For salads I much prefer the flavor of Red Russian to green Siberian. And for salads I prefer *B. napa* to *B. oleracea*.

Collards are non-heading cabbages that are grown like kale and are nearly as tolerant to moisture stress as kale. Unlike kale, the leaves taste sweet during hot weather. Collards thrive during winter in the Deep South. Collard greens are much appreciated in the southern states.

Kohlrabi (Regular and Giant)

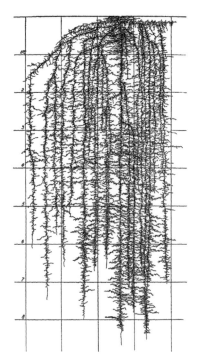

Mature giant kohlrabi

Irrigation may not be required if the spring sown kohlrabi crop is given elbow room. With market gardener varieties, try thinning to 5 inches apart in rows 2 feet apart and harvest by thinning alternate plants from a row. Small kohlrabi varieties grow fast and mature quickly, hybrids especially do this. The spring sowing must get an early start and be abundantly fertilized to grow fast enough to be harvested before hot weather makes them get woody. In Cascadia on irrigated ground sow another patch in the first half of August for autumn/winter harvest. In the humid east sow the late crop about two months before autumn begins. In the Mid South sow an autumn patch in mid-August. In the Deep South sow mid-August through October and again in early spring. Keep in mind that slow growth makes kohlrabi get fibrous and lose sweetness. Keep them growing faster by harvesting alternate plants in the row.

Kohlrabi is also used as animal fodder in Europe. These varieties were bred to grow to soccer-ball size, like rutabagas. Fodder types taste good to livestock but some were later refined for human tastes and became "giant" vegetables that suit dry gardening far better than refined market types.

What to do with a giant (or regular) kohlrabi? Peel, grate, add chopped onion, dress with olive oil and black pepper, toss, and enjoy this old Eastern European mainstay.

In Cascadian dry gardens try sowing giant varieties late in April; hope for a foot-tall plant before really hot weather arrives. Make positions 3 feet apart in rows 4 feet apart. If they get fertigation during summer they will grow as large as soccer balls. This approach probably works in cooler summer areas of the humid east. It should work fine too in the Deep South for a cool season crop started September/mid-October.

I discovered Superschmeltz in a Swiss seed catalog when I ran Territorial Seed Company. Territorial still sells it and says of it: "Other giant kohlrabi types tested by Territorial over the years have proven to be woody and throw a high percentage of off-types. Superschmelz is the only giant variety we've tested that grows uniformly large and remains very sweet and tender to the biggest sizes. It produces juicy 8–10 inch bulbs, and *the large root system lends itself to dry gardening.*" Johnny's Kossak is a new hybrid variety that should do as well and likely is more uniform than Superschmelz.

Leek

In Cascadia leeks can serve the water-short gardener as a dependable winter-hardy onion substitute that supplies the kitchen from late summer until mid-spring the next year. Ordinary bulbing onions can not be grown in Cascadia without plentiful irrigation. Onions do succeed without irrigation in the humid eastern states; there leeks hold less attraction because this vegetable does not survive in the garden after the soil freezes; leeks do not store as well as onions after harvest.

Leeks take a long time to get large, even when growing conditions are excellent. Given poor conditions they never do get large. *Young leek plants have a weak, inefficient root system that requires fertile and constantly moist soil for the first few months.* The crop gets a much better start if leek seeds are sown in an always airy, always moist super-fertile nursery bed made in the garden, *not in containers.* If you search the net for leek growing instructions you will always be told to start the seeds in flats, pots or small trays and transplant when they are three to four inches tall. After transplanting these spindly seedlings require constantly moist soil until their stalks are at least the diameter of a pencil. You may save a lot of water by confining essential irrigation to a compact nursery bed until their stalks are pencil size before transplanting into their final position. In the humid eastern states locate the nursery in soil under a cold frame, under a cloche or under a polytunnel, not in containers on a greenhouse bench. Sow as early in spring as germination can be obtained. Extension service publications from the humid eastern states say to start February/March. Cascadians can start transplants in an unprotected nursery, sowing late March to mid April depending on the microclimate. In the Middle- and Deep South start the seedling nursery as early in September as possible, immediately after the heat of summer breaks.

Spread an inch thick layer of compost and complete fertilizer over the nursery bed and dig them in at least 6 inches deep. This creates a highly favorable root zone. Here in Tasmania my leek nursery has been 4 feet by 4 feet, with three rows of seedlings in it. It is located close to the garden hose tap because every time I use the hose the nursery also gets a brief shower. (The celery patch is also located close by.) That little nursery bed produced three times as many transplants as the most serious gardener would ever need, but I like to gift friends and fellow garden club members with fat bundles of leek seedlings. Thin the row to 1/8th inch apart when the seedlings are 3 inches tall. The seedlings should compete strongly enough that they form long shanks (the above ground part of the plant below the lowest leaf). Irrigate the nursery as often and as much as necessary to keep the surface inches comfortably moist at all times. I top-dress a little extra fertilizer close to the row whenever growth slows; I do this for the last time when the seedling are around 8 inches tall. I urge you to do that too, but be careful of overfertilization, which will work against you. When topdressing, less more often is better than more less often.

Transplant when the shanks are pencil-thick. Make sure the top 6 inches of the nursery is moist. Then dig the seedlings with a spade, leaving 4 to 5 inches of soil still attached to the roots. If enough compost had been mixed into the nursery bed before sowing and if that soil is quite damp it should be easy to shake the soil loose from the roots without causing too much root loss; or else blast away the soil with a strong jet of water as you gently separate the tangled roots. Break off as few roots as possible. After untangling the seedlings I snip off one- third to one-half the photosynthetic leaf area (not counting the shank) to reduce transplanting shock.

If I am not going to transplant immediately I temporarily bundle the seedlings in wet newspaper to make sure the roots don't dry out. I place these bundles in a plastic bag and hold them in the fridge for up to a few days before gifting them or transplanting them myself. If I am going to transplant that same day I put the seedlings in a bucket, roots down, and add an eighth-inch of water to the bottom, not enough to submerge the roots, but enough to keep them very damp for a few hours while I stash that bucket in the shade. If there are not enough seedlings to loosely fill the bucket and hold themselves upright, I sprinkle damp loose soil over the roots, of which I have a'plenty from just having shaken it off those roots.

The seedlings are transplanted single file down the middle of a three- to four foot wide bed that is not routinely walked on. Dig an 8-inch-deep trench the width of an ordinary shovel down the center of the bed and temporarily place the earth you removed next to the trench. Cover the bottom of the trench with a heavy dose of complete fertilizer and a half-inch layer of compost and spade all that in so the soil is fluffy and fertile at least another 6 inches down. Using the end of a shovel handle or a dibble, poke a row of 6-inch-deep holes along the center of the trench.

The amount of separation between plants depends on how much rainfall you anticipate and/or the amount of irrigation possible. In Cascadia if there is no irrigation possible, set them 12 inches apart. In the humid eastern states give them 6 inches of separation. With unlimited irrigation 2 inches between plants is enough to harvest really big ones.

If the nursery bed produced excellent seedlings there should be about 4 inches of pencil-thick shank on each seedling below where the first leaf attaches. Place one leek seedling in each hole and set it deeply enough that the lowest leaf notch is barely exposed, then mud the roots in with a cup of liquid fertilizer. Then spread additional fertilizer on the bottom of the trench before any more soil is pulled back into it. What I'm getting at here is that you want to GROW these seedlings as rapidly as possible! As the leeks gain size gradually pull soil back into the trench using a common hoe. When the trench is full, pull up more soil from both sides of the bed against the stems, making a ridge that eventually may stand as much as 6 inches above ground level. This practice resembles hilling up potatoes. Light deprivation makes the best-tasting white part of the shank become as long as possible. When pulling soil around them avoid getting any in leaf notches or you'll not get them clean after harvest.

In the humid eastern states winter gets too cold for leeks to survive in the ground. Dig the crop before the soil freezes and store the leeks under cool (32°F to 40°F) and moist conditions, best packed upright shoulder to shoulder in boxes of moist peat or sand with the leaves uncovered and stored in a basement or root cellar. Where winter is not severe it may work to mulch them heavily in autumn, digging them as needed from soil that is not quite frozen. Once the leeks freeze they will be ruined so if nothing else, a thick straw mulch may lengthen the harvest window.

In the Middle- and Deep South start and GROW them with all your ability in an irrigated nursery bed after the summer's heat breaks. After transplanting out into the garden they will do fine without irrigation. In December they should be large enough to begin to harvest. Dig them until they go to seed in spring. Gardeners there will have to work out the best between-plant spacing based on experience. My best guess is 3 inches between plants.

Lettuce

Lettuce forms a deeply penetrating moisture-storing tap root that allows the plant to survive moisture stress, yet lettuce will not taste good if it's short on water. It should be clear from the root system drawing that one lettuce plant is capable of extracting all available moisture from a three foot diameter circle.

In Cascadia lettuce sown early- and mid-spring will grow large and be sweet and tender without irrigation if the plants are positioned 1 foot apart in a single row with 18 inches of un-contested elbow room on each side. Cascadian lettuce maturing after mid-June without regular, generous irrigation will taste bitter. Large raw greens salads are so important to my health that I'd use precious water to irrigate summer lettuce. Water-short Cascadians can start the fall harvest on an irrigated nursery bed mid-August and transplant it out after the fall rains return. It would help if this nursery were protected by shade cloth because lettuce seed will not germinate in hot soil. This is also the case in the Middle South. In the Deep South lettuce is a cool season crop; if given plenty of growing space it should not require any irrigation.

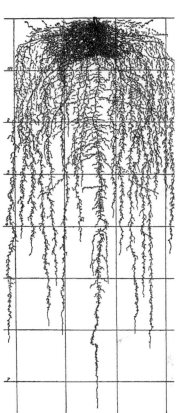

Lettuce at maturity

Melons

Everywhere in Cascadia north of Roseberg, Oregon is barely warm enough to ripen a few fruit on the most chill tolerant melon varieties. Cascadians will find these cultivars offered by Territorial and West Coast

Seeds. To achieve any ripe melons in most of Cascadia the crop needs to take advantage of every warm summer day. Sow seeds as soon as they'll germinate outdoors—at Elkton, Oregon that is May 15 to June 1. In the Willamette Valley that can be June 1 to June 15. See "Cucumbers" for a way to pre-sprout (chit) the seed that makes germination much more certain. Sow three seeds at each position. Thin to a single plant when there are about three true leaves and the vines are beginning to run.

In the humid eastern states, even in much of the northern tier of states and in Canada close to the US border there is plenty of summertime heat. No need to rush. Start the fast-growing vines by mid-June. In the Deep South start them late March through July. In the Middle South, May through mid-June. Most varieties will cover a circle 8 feet in diameter. Without irrigation, space the hills 8 feet apart in all directions. The root system fills at least as much soil as the vine covers. Fertigation

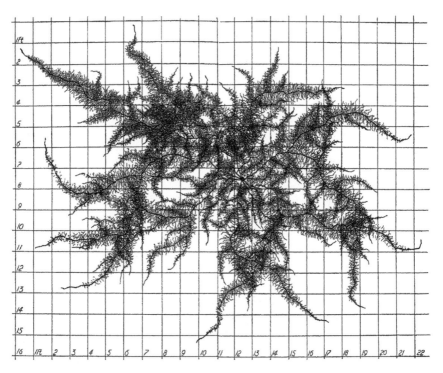

Melon root system at maturity

every two to three weeks will increase the yield by two or three fold while making the melons larger and possibly sweeter. Release at least half of the water/fertilizer mix close to the center of the vine, where the taproot can use it.

Onions /Scallions

In Cascadia spring-sown bulbing onions and scallions require abundant irrigation. But the water-short, water-wise Cascadian can still supply the kitchen by growing overwintering bulb onions and leeks. Full-sized leeks can be harvested from October through early April. Overwintered bulb onions are harvested late in spring and store through the summer. Scallions may also be started in a nursery midsummer and transplanted out at summer's end when the rains start. These are harvested during winter. Territorial and West Coast sell productive overwintering onion varieties that are adapted to northern daylengths and cold hardy enough to survive winter there.

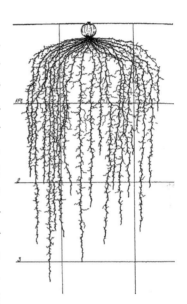

Mature onion

Getting the sowing date right for overwintering onions is critical in Cascadia. If they are sown before mid-August they will grow too large before winter. After their stalks are more than lead pencil thick, periods of harsh weather trigger them to form seed next spring. You do not want onion seeds. You want bulbs. To achieve that in Cascadia the shanks of overwintering onions must be no more than 1/4 inch thick by December. However, if seeds are sown too late the spindly plants may not survive winter and will for sure produce smaller onions. Most gardeners have to experiment a few times before they learn the ideal sowing date for their location. The first time you try it I advise sowing August 15.

Gardeners with plenty of irrigation water can directly seed them. Water-short gardeners should sow the seed in an always moist and very fertile nursery bed; you want to end up with one seedling every eighth-inch in the row. If you start more than one nursery row, separate them by 12 inches. Read more about nursery beds in the section about growing leeks. These seedlings should be transplanted out by mid-November. Those who miss this window of opportunity can start a transplant nursery in early October and cover it with a cloche. This accelerates seedling growth during fall and early winter. These seedlings should be dug and transplanted to their final position as early in spring as it is possible to fertilize and turn over their bed. Soil moisture will be of no concern when the plants are growing fast and then forming bulbs. Transplant the seedlings 3 inches apart in rows 16 inches apart across a wide raised bed.

Top dress directly seeded overwintering bulbing onions when growth accelerates in spring. This will result in much larger bulbs.

I've found I get the best growth and largest bulbs when onions follow potatoes. After the potatoes are dug in early October, immediately prepare the bed to receive onion transplants (or garlic cloves).

Water-short Cascadian gardeners may sow scallions in a nursery just like overwintered onions, but start them mid-July so they'll be large enough for the table during winter. Transplant them 1/2 inch apart in rows 12-to 18 inches apart. Scallions will go to seed March/April.

In the humid eastern states bulbing onions usually do fine without irrigation if they are not crowded. Direct seed them in spring while the topsoil remains moist; the seed germinates well in cool soil. Thin seedlings 3 inches apart in the row; make the rows 18 inches apart. In the Deep South grow overwintering onions much as it is done in Cascadia, but use locally appropriate varieties, often termed "short day" onions. Sow them late September and before mid-October. In the Deep South there is little risk of overwintering onions bolting to seed next spring because winter conditions are not harsh. In the Middle South both spring sown varieties and the very cold hardy varieties of overwintering onions used in Cascadia can be grown.

Parsley

In the humid eastern states and in Cascadia sow parsley seed early enough in spring that the soil naturally remains moist until the seeds germinate (takes 2 to 3 weeks) or else irrigate a narrow band of soil where the seeds rest to keep it moist. By the time the seedlings are two inches tall thin them so they stand 6 inches apart in a single row that owns a space 4 feet wide. When these plants bump into each other, thin again so they stand 18 inches apart in the row. Given this much growing room these plants will get big. Five plants might supply the average kitchen. If you direct seed their tap roots remain functional; in that case irrigation should not be necessary unless yield falls off too much during summer, and that is unlikely. Parsley's very deep, foraging root system resembles that of its relative, the carrot. If you transplant parsley seedlings the tap root gets broken and the parsley patch may need fertigation.

In the Deep South parsley is best considered a cool season crop. Sow it after the summer heat breaks. Parsley produces during winter and spring but goes to seed later in spring. In the Middle South sow parsley in spring and again in August; it will probably overwinter.

I never grew root parsley without irrigation. I imagine it will do as well as leaf varieties. Parsley root tastes much like parsnip; however, parsnip is far more likely to handle dry soil and is far more productive than root parsley.

Parsnip

Grow parsnip like carrot. The further apart they are in the row and the more between row space they are given, the better they'll handle dryish soil. Parsnips are primarily subsoil feeders; after their leaves are 4 inches tall it doesn't matter much if the topsoil gets dryish as long as there is adequate subsoil moisture. With full irrigation I make rows 18" apart and thin the plants in the row to 2- to 3 inches apart. If you can count on rainfall keeping the soil reasonably moist during summer thin the plants to 3- to 4 inches apart in rows 20 inches apart. If summertime moisture stress is anticipated, make your rows 36" apart and thin the plants in the rows to 6- to 12 inches apart. Parsnip is slower than carrot to form big roots.

Pea

Peas are grown in spring and early summer when the soil starts out moist. The crop usually matures without irrigation if the bed is not crowded. Irrigated or not irrigated, both pole and bush varieties should be thinned in the row to an inch apart; however, make unirrigated rows 3 feet apart. Alaska peas harvested as dry seed for soup take the same treatment. In Cascadia, in the East and in the Mid South sow the seed as early in spring as strong germination can be obtained. In the Deep South sow peas anytime between early September through January.

Southern peas/cow peas, are a heat loving bean that cannot take frost. They do not do well in cooler climates but thrive in the Deep South, where they are much appreciated as a green shell bean as well as a dry bean. Southern peas can be planted when the soil has warmed up in the spring and can be successively planted through the hot summer months. They are good at growing in poor soil and work well as a combination nitrogen-fixing cover crop and vegetable to keep unused beds occupied during the heat of a Deep South summer. If summers are very wet, the Southern peas may not work as dry beans because they tend to rot and mold in the shell rather than drying down.

Pepper

When pepper plants are spaced the usually recommended 16 inches apart on raised beds they experience intense root zone competition before their leaves have even formed a complete canopy. These crowded plants require plenty of irrigation. The plants grow much larger and bear much more- and larger fruit if they are given plenty of elbow room. Set out transplants at the usual time for your region. For each plant first spread a half inch thick layer of compost plus complete fertilizer over a 12 inch diameter spot and deeply dig them in. Position these spots 3 feet apart in rows 3 to 4 feet apart. For an abundant harvest, fertigate every 3 weeks.

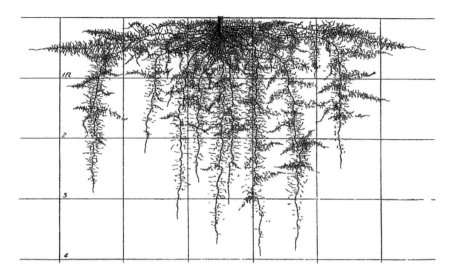

Mature pepper

To achieve the biggest pepper plants you ever grew, fertigate generously every 7 to 14 days—how frequently depends on the depth and moisture-holding capacity of the soil.

When I grew peppers in Cascadia entirely without irrigation or fertigation only small-fruited varieties were productive. Small-fruited types, both hot and sweet, have much more aggressive root systems and generally adapt better to Cascadia's cool summer nights. I urge Cascadians to grow pepper varieties sold by Territorial, Adaptive or West Coast Seeds. In the Deep South, hot peppers thrive in the heat and humidity and outlive bell peppers, often producing all the way through summer and fall until killed by frost in winter. In regions without frost, some hot peppers, such as Thai, habenero and cayennes, will live as large perennial shrubs for three years or longer.

Potato, Irish

The potato produces an enormous amount of food per acre. It rivals rice in that respect. Even after deducting its 90 percent moisture content, the potato still produces more food per acre than wheat does. If the soil provides the full range of plant nutrients in balanced abundance then the potatoes will be highly nutritious. The potato may contain up to 11 percent protein. The supermarket potato commonly provides only 8 percent protein. That roughly 30 percent difference can make all the difference to the health and longevity of people who use potatoes as their staff of life. The 11 percent protein potato also tastes really good all by itself, or at most with a bit of butter. It does not need ketchup, melted cheese, etc.

Kale nutritionally compliments potato. Please read in the section about kale (again) what I say about mashing kale and potato together to make colcannon. And recall my assertion that kale produces more high-quality protein per acre than any legume crop. Both these vegetables are among the easiest to dry garden.

The wild potato thrives in the cool, arid high plateaus of the Andes where annual rainfall averages 8 to 12 inches. Potato makes an extensive root system that can find enough moisture without being

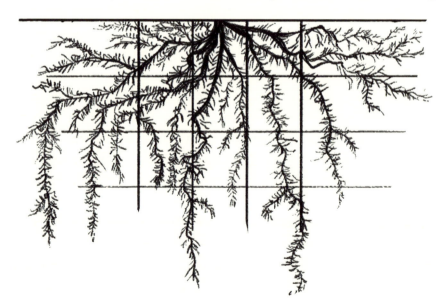

Potato root system at maturity

irrigated if the grower uses a low plant density. If moisture stress is likely then choose a layout such that when the vines have reached their full extent there will be a narrow foot path to either side of the row—in short, four- to five feet between row centers. Vine length is partly determined by variety and partly by how much virus disease the seed carries. When I have grown certified varieties and virus-riddled heirlooms in the same row I have seen vine length vary from 18 inches on the heirlooms to over 30 inches. I assume root extension matches vine length. Certified virus free seed is much much more vigorous—and much more productive. Certified seed is well worth the price!

In the early pre-irrigation days of the Idaho potato industry (where annual rainfall is 10- to 15 inches) growers positioned seed pieces 18 inches apart in rows 5 feet apart. You can read about this in Widstoe's *Dry Farming.* * These days irrigation is always used in Idaho; it greatly increases yield. However, when the same variety grows under moderate moisture stress in truly fertile soil it can contain up to 30 percent more protein and far more nutrition of every sort, which makes it taste better. Perhaps the high quality achieved in the early days is why "Idaho" and "potato" are now so closely associated.

East of the 98th meridian (a line on the map going north from Dallas, Texas), most years rainfall supplies the potato crop with plenty of moisture. I have to warn you: Using 18-inch in-the-row spacing and far apart rows in fertile soil that supplies plenty of moisture results in harvesting mostly *very large* potatoes but does not necessarily result in harvesting that much less overall weight from the area involved had the grower used a higher plant density. If you want the potatoes to be smaller and are confident there will be no moisture stress then position seed pieces 10 to 12 inches apart and make the rows 4 feet apart. If rainless periods lasting weeks are possible consider using 4 feet between rows and 14 inches between seed pieces.

In Tasmania, where I live now, summers are nearly as dry as they are in Cascadia. Before irrigation became commonplace on Tassie, dry-farmed potatoes were an important export crop. Tasmanian commercial growers these days irrigate generously and use 8 inches between seed pieces, the rows are a scant 3 feet apart in order to have most of their harvest meet supermarket- and processor size requirements.

Between row spacing determines how weeding will be done. If rows are only 3 feet apart it soon becomes impossible to hoe- or walk between them. Spraying herbicide "solved" this problem. But gardeners do not poison weeds. We hoe them. Or hand pull them. Or we thickly mulch the potato bed. Mulching works when the family's entire potato crop occupies no more than 100 square feet but I think hoeing weeds while hilling up works better when the family seeks to put away enough potatoes to subsist upon them from the end of summer until the middle of next spring. Gardeners should provide enough between-row space that they can wield a hoe while walking between rows.

When I dry garden spuds I plant one seed piece every 18 inches down the center of a 4-foot-wide slightly raised bed with an 18 inch wide path between beds. This plant density is similar to what Idaho dry farmers did in 1900. If I plan to irrigate I use the same between-row space but position the seed pieces 12 inches apart.

My wife and I do not make potato our staff of life; in the interest of achieving greater health and longevity we keep carbohydrate foods to a minimum in our diet. We do put a small serving of potato on our plates, maybe two times a week. In order to generously supply the two of us I allocate two beds that are four feet wide and 25 feet long. Half of one of those beds grows the early crop that we eat during summer. If I were counting on potatoes to fill our belly several times a day from July through April I'd grow at least three times that amount. And that's just for two oldies who do not eat nearly as much as we once did.

Now for the growing details. Spread fertilizer and a quarter inch thick layer of compost over a four foot wide bed. Dig them in or rototill. Place the seed pieces only 2 inches below the surface in a row down the bed's center. This is less deep than most garden books advise and less deep than farmers do these days. But farmers no longer hill up their rows. They now plant deep enough to avoid ending up with too many green potatoes and use herbicides to control weeds. Here's how to hill up. When the first vines to emerge are only 4 inches tall, stand on the path, reach over the bed as far as you can with a common hoe, press the blade lightly on the earth and pull in, depositing about one inch of soil up against the vines. Then repeat that action while standing on the other side of the row. Let them grow another 4 inches and then scrape up another inch of loose soil over what you hilled up before. Continue doing this about once a week until the vines have grown so numerous and so long that hilling up can no longer be done without causing damage. At that point there should be a row of blooming potato vines emerging from a mound of loose soil 12 to 16 inches high and about the same width.

Hilling up simultaneously eliminates weeds and forms a dust mulch. It also insures that the soil around the forming tubers is loose and airy. This is essential! High yields do not happen when the soil is tight and resists tuber expansion. Before herbicides farmers routinely hilled up; in that era they could grow potato in soils with some clay content. Now they plant seed pieces about 6 inches deep and do not hill up so they can only grow spuds on sandy or light loam soils that naturally remain loose. Should an odd weed emerge through the vines, pull it by hand. It'll pull out easily because the hilled up soil is not compacted and it has been protected from being compacted by rain or overhead irrigation by the vines' leaf cover.

It is essential that developing potatoes do not get exposed to sunlight or else they turn green and become poisonous. *Do not eat green potatoes.* Do not think you can peel off the green bits and eat the rest. Do not feed green potatoes to livestock. A green potato contains poisonous solanine throughout. If you get more than a few green ones, then next time you grow potato scrape up more soil over the row.

You will grow potatoes better with an appreciation of the plant's strategy. The leaf is their sugar factory. Initially this sugar is reinvested to grow more leaves as fast as possible. Leaves contain a lot of protein.

Manufacturing protein at a high rate requires plenty of nitrogen. After about a month of forming new vines and roots the plant begins blooming. At the same time flowers appear the plant starts forming potatoes; they are attached to the vines below the soil line and *above the seed piece*. Initially they are just tiny bumps the size of garden pea seeds. During the blooming period new vines keep emerging and new potatoes continue to be set. The plant continues to need lots of nitrogen so it can be smart to foliar feed vines every week to 10 days; I opine this practice can increase yield by half. Then blooming stops and with that the plant stops growing new vines; the initiation of new potatoes stops as well. Now the plant concentrates on storing starch and all the potatoes get larger. Making starch out of sugar does not use nearly as many plant nutrients as growing new leaves does. Serendipitously, at this stage the fertility you invested in the potato bed has largely been used up. Foliar feeding at this stage doesn't help much. I stop doing it. As the existing leaves age their photosynthetic efficiency declines, eventually the leaves brown off, then the potatoes stop swelling and form tough skins that withstand long storage.

In summary: Push the plants to grow as fast as possible from the get-go until blooming ceases. During the time the potatoes are swelling continue to hand pull any weed that emerges through the vines or competes with them for soil moisture. Footpaths along the row should be cleaned of weeds. When the vines start deteriorating you can stop weeding unless a weed begins to form seeds. Any weeds that do get going among the potato vines will be killed when you harvest. As the vines deteriorate and brown off naturally, the tubers toughen their skins, which leads to much better storage. Ideally they are dug soon after the vines have completely died back. If an early frost should kill the vines before they have died back by themselves, allow the spuds to remain in the ground for a few weeks while their skins get thicker and stronger. Then the potatoes won't be damaged when you dig them.

When I lived in Cascadia I started two potato patches every year. The first was ritually done on St. Patrick's Day—March 17th. Rain or shine, be it in untilled mud or finely worked and deeply fluffed earth, I still planted a dozen seed potatoes of a quick maturing variety. This planting provided our kitchen from July through September. Potato vines are not frost hardy but at Elkton, Oregon there was considerable frost risk until mid-April so I covered the vines with spun fiber netting that protected them down to maybe 28°F. The row cover also accelerated early vine growth and hastened the formation of potatoes. I would remove the covering in order to hill up and then reposition it. After a few repeats the row cover would be torn and tattered; there was no more frost risk so it was discarded. Planting on March 17th is definitely too early in the Willamette Valley and north of there. In the more northern parts of the eastern states shaded patches of snow may still be melting on St. Patrick's Day. A better way to determine when to attempt a frost-protected early crop would be after the daffodils are in full bloom. March 17th is about the latest date to use in the Deep South; it is pretty spot-on for the first planting in the Middle South.

In the humid eastern states planting the summer's supply should wait until frost is highly unlikely and for the soil to have dried down enough that you can spade in fertilizer/compost 12 inches deep over a four foot wide strip. Do everything possible to preserve subsoil moisture before planting. Keep the prospective potato bed and surrounding paths entirely weed free starting early in spring. There is no rush to start the main crop except perhaps in short season cool areas close to the Canadian border. The potatoes will keep longer in storage if the planting date you choose causes the vines to die back in the last weeks of summer, not sooner than that.

No farmer would attempt to grow potatoes for profit on clay soil but you are a gardener, so if clay is your situation you want to encourage the surface foot to dry down enough that you can dig a crumbly seed bed. Letting the spring weeds grow unmolested helps to accomplish that—as long as you eliminate them before seeds form. Growing a spring green manure crop like peas helps accomplish the same. If none of this seems possible, then this might be an occasion to grow potatoes under mulch instead of hilling up soil. Plant the seeds just below the surface and cover them with an inch of loose straw. As the vines grow scatter loose straw such that the vines can get through it. Potatoes will form on the soil's surface; the straw will protect the majority of them from greening up.

In the Deep South, Irish potatoes can be planted from late January through February so they finish before the full heat and humidity of summer. Potatoes can be planted again late August/early September to be dug as needed during the cool season. In the Middle South they are planted March through mid-April for summer harvest and again in July for autumn harvest. If you wish, you could grow a cool-season spring crop of something fast like dwarf shelling peas in early spring, then harvest that and plant your potatoes in July to get two vegetables from the same space in a single growing season.

There is no need to do anything to prevent cut seed potato pieces from rotting instead of sprouting if you'll green up the seed potatoes and get them sprouting before planting them. The British call this "chitting," a procedure similar to what I suggested a few pages back for cucurbit seeds. Spread out *whole uncut* seed potatoes in a bright warmish room for two to three weeks before planting them. Or do this in a garage or outbuilding where temperatures will be lower but the structure must have a window. In this case start a month before it is time to plant the seed. There must be enough light to green up the seed, which will also trigger sprouting.

Plant whole sprouting small potatoes that weigh between 1½ and 2 ounces (called single drops in the industry) or else plant cut pieces weighing at least 2 ounces that contain three eyes. Three eyes is important. More than three and there are too many vines emerging from each position, and too much competition. Fewer than three will lead to very large potatoes, and possibly less overall weight harvested. There is a lot of garden lore about how to prevent cut seed pieces from rotting instead of sprouting. Chitting makes these precautions unnecessary. I suggest cutting seed potato pieces immediately before planting them. Do not let them dry out. There is no need to let the cuts skin over. If the seed potatoes had been chitted and are as large as I suggested they will not rot in the ground. You are a gardener. You can handle sprouting potatoes gently; take care not to break off any shoots.

My main potato difficulty in Cascadia was too-early maturity. Early fast-finishing varieties don't store well unless they brown off late in September, not in August. Potatoes only keep in storage for a long time when they are kept very cool, entirely without light and in a very humid environment with good air exchange—conditions almost impossible to approach on the homestead during summer. The best August compromise is to leave mature potatoes un-dug until the beginning of autumn. However they'll likely already be sprouting in the ground by the end of September. We found this was acceptable for the odd unused bits of my St. Pat's Day sowing. We just dug them, broke off the sprouts, put them in a cardboard box to keep them in the dark and ate them first. Already sprouting at harvest time is not okay for the main winter storage crop. Late-maturing varieties usually yield more.

Potato, Sweet

Only this segment of my book is not written from personal experience, but absolutely this information did not come from garden books or magazine articles. I entirely gave up reading that pabulum when I went into the seed business in the early 1980s. While writing the new edition of this book I confess that I did consult extension service publications from the Deep- and Middle South regarding vegetable crop planting schedules and of course, I considered rainfall and temperature statistics. I found writing about growing sweet potato particularly challenging because I have never attempted to grow it; however, the gaps in my knowledge have been filled in from extension service publications flavored with my own understanding of how plants grow, as well as additional knowledge from David The Good's experience with this tropical vegetable.

Sweet potato is adapted to a warm and humid environment, both above the soil and in it. Moisture stress greatly lowers yield, size and quality. Waterlogged soil is not good either. Commercial growers in the South consider irrigation to be mandatory and are quite specific about the soil moisture levels they intend to maintain. Growing them without irrigation leaves the grower with the possibility of achieving a low yield of commercially unmarketable tubers.

Sweet potato requires at least 1 inch of water per week to grow really well. Lowered plant density is how the dry gardener can compensate for a temporary rainfall shortage. Louisiana State University says home gardeners should space the plants 12 to 14 inches apart in the row, and between row spaces at least 3 to 4 feet. If I were dry gardening in the South I'd try 18 inches apart in the row. All the official publications advise building a low ridge on which to transplant, which helps to prevent waterlogging when there are heavy summer rains. Especially do this on soils containing a fair bit of clay. In well-draining sand making a mound is not necessary. Till the soil and then form flat-topped ridges about 10 inches wide and 10 inches high, spaced about 4 feet apart center to center. Transplant the shoots 12 inches to 14 inches apart on those ridges. Avoiding dry soil is essential while these shoots are first forming roots. Gardeners without enough water to irrigate the whole patch could spot water these positions until the vines start running. In order to preserve all soil moisture for the crop weeds must be thoroughly eliminated. The official publications assert that once the vines begin to run they usually will shade out the weeds. This is generally the case, as sweet potatoes are very aggressive ground covers.

Gardeners in the North also grow sweet potato. They can succeed, especially when they use an early-maturing variety like Georgia Jet. However, northerners can not make the sweet potato their "staff of life" because they can not conveniently store a large harvest for a long time. Long storage only happens when conditions are warm and dry, such as inside a heated house. Northerners are far better off to depend on the Irish potato. Despite the climate's cool nights, Cascadians sometimes succeed at harvesting a few sweet potatoes. I don't care much for eating them so I never tried it. In the Deep South, there are both sweet and starchy varieties of sweet potatoes that thrive despite less-than-ideal soil, heat, torrential rain, humidity and bugs. In portions of Florida white potatoes may rot and fail while sweet potatoes sail on through.

Rutabaga

This brassica makes a useful low carbohydrate alternative to potatoes, which should interest people on a keto diet. They can keep as long as potatoes under more or less the same conditions potatoes are stored at. The yield per space involved can be huge.

Rutabagas have wonderfully aggressive root systems and are capable of growing slowly through long rainless periods. But . . . when doing dry gardening experiments in Oregon I started rutabagas early in April when seed germination was easy to achieve. I positioned them 3 feet apart in rows 4 feet apart. The plants grew steadily through the rainless summer. By October they had formed basketball-sized hollowed-out shells that had been wrecked by cabbage root maggots. In Cascadia root maggot levels are at their peak in the month of May. Most of the damage they cause can be avoided (in Oregon, but maybe not in Washington state) by sowing brassica crops mid-June through July—a time when topsoil moisture has been severely depleted. One workable strategy would be to preserve soil moisture by spreading a straw mulch over the prospective rutabaga patch late in April, pull it back and expose a two inch wide strip of soil just before seeds are to be sown late in June, spot water those narrow ribbons until germination occurs and then gently coax the mulch back around the (tiny) seedlings.

In the last week of June, 1991, we got two very surprising inches of rain. So as a test I sowed rutabaga on July 1. The seeds germinated without irrigation. With no rain for the rest of the summer they became somewhat stunted. By October 1 the tops were still small and looked moisture stressed; the edible bottoms had not developed much. Then the usual Cascadian rains came and the rutabagas began growing rapidly. By November there was a pretty nice crop of medium-size, good-eating roots. So I speculate: If the seed were sown July 1 and a little irrigation could be applied with a hand nozzle close to the row then a decent crop might be had most years.

In the Middle South and Deep South, start rutabaga in August. They probably will not need any irrigation, except possibly to help germinate the seeds.

Sorrel

This drought-tolerant perennial salad weed is little-known and underappreciated. Leaves forming in the heat of summer have an objectionably strong flavor. That's okay; there's plenty of other stuff to eat then. Dur-

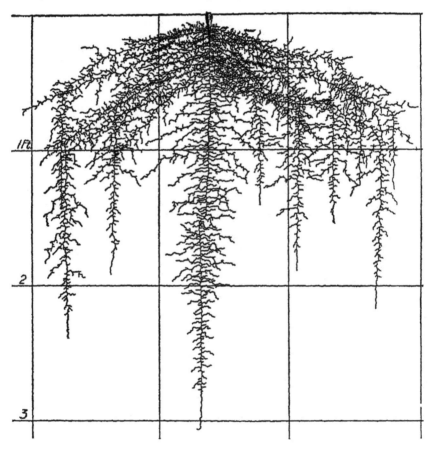

Half-grown rutabaga

ing cool weather, sorrel's lemony taste and tender texture balance tougher savoy cabbage and kale and turn them into much more interesting salads. Our kitchen has used most of the cool season production from 4 row-feet.

In Cascadia the first year you grow sorrel, sow mid-March to mid-April. The tiny seed must be placed just below the surface. It only sprouts when the surface stays moist. Sprinkle the seeds into a ¼ inch deep furrow scratched with the corner of a common hoe, the furrow centered in a space 4 feet wide. Do not cover the seeds. Water them immediately with a hose/nozzle, which washes enough soil into the furrow to barely cover most of the seeds. Water again every time the soil's surface dries out until germination occurs. After that no further irrigation should be needed. As the seedlings grow, thin gradually. When the leaves are about the size of ordinary spinach, individual plants should be about 6 inches apart in the row. In summer the taste is so strong you won't eat it so there's no point in pushing sorrel to grow fast. If production lags in fall, winter, or spring, side-dress the sorrel patch with compost or organic fertilizer.

Sorrel is perennial in Cascadia; it may survive winter as far north as USDA Plant Hardiness Zone 5. If an unusually harsh spell of frosty weather kills off the leaves, it will probably come back in early spring. In Oregon I have seen sorrel survive a few consecutive days when temperatures at sunrise were 3°F and the soil froze an inch deep. In spring of its second year and thereafter, sorrel will make seed. Seed making saps the plant's energy and the seeds may naturalize into an unwanted weed around the garden. In the South, sorrel is already a common weed of horse pastures and unkempt lawns. So, before any seed forms, cut off all the seed stalks and use them to conveniently mulch the row. If you move the garden or want to relocate the patch, there is no need to start sorrel again from seed. In any season dig up a few plants, divide the root masses, trim off most of the leaves to reduce transplanting shock, and transplant 1 foot apart. Occasionally a unique plant will be far more reluctant to make seed stalks than most others. Making seed is an undesirable trait so I propagate only seed-shy plants by root cuttings.

Spinach

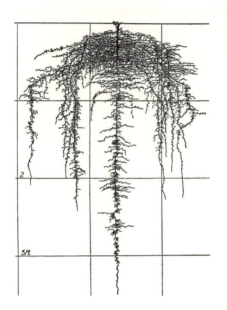

Ten week old spinach

Spinach tolerates moisture stress far better than it might seem if judged by its delicate structure and the succulence of its leaves. In the entirely rainless summers of Cascadia without any watering at all a bolt-resistant, long-day variety bred for summer harvest that was sown in late April may still yield pickable leaves in late June or even early July. Thin the row to 12 inches apart; make rows 3 feet apart. Grown that way the plants get a lot bigger than you might expect. So do the leaves.

If you establish a lower than usual plant population, unirrigated spinach should do fine in the Middle South both spring sown and again late in summer. In the Deep South start spinach October through February.

Squash, Winter and Summer

This family's root system can be far greater than most people realize. When cucurbit vegetables are directly seeded (not transplanted) a deep taproot extracts subsoil moisture. Soil conditions permitting, shallow feeder roots extend laterally as far as or farther than the vines reach at their greatest extent.

Dry gardeners can do several things to assist cucurbits. First, at the time you sow seeds make sure there is absolutely nothing growing in their entire potential root zone, including weeds. Where there is no squash vine borer grow only one plant per hill with the hills separated in all directions a little farther apart than the greatest possible extent of the variety's ultimate growth. Where one squash vine borer can destroy an entire plant, it is safer to start with 2 plants at each position. Common garden lore states that squash vines droop their leaves in midsummer heat, that temporary wilting cannot be avoided and does no harm. But if they're grown on deep, open soil and given as much space as I just described then hot afternoon wilting will be unlikely even if there is no watering. Two plants per hill do compete and both may wilt when the sun is strong.

Second, deeply dig and fertilize the entire potential root zone. It is best to do this in stages, starting with the planting hill. When the vines start running fertilize and dig concentric rings that are two feet wide; do this one week before the rapidly expanding vine reaches that far. Going at it this way makes the planting the summer garden less stressful in that the entire squash patch does not have to be dug when there is so much other garden work to do. Again, do not transplant them. This breaks the taproot and makes the plant more dependent on lateral roots seeking topsoil moisture.

Delaying sowing until the soil is warm ensures germination and rapid early vine growth. A cucurbit that gets a stressless start will soon outgrow one that barely emerged a few weeks sooner and struggled. It's best to wait for warm weather before sowing—and chit your seed. See the section on cucumber for information about chitting.

Some locations alternate between warm and cool in the spring and a short pleasant growing season that rapidly gets too hot. If you aren't sure whether the time is right in your area and you want to get a head start on spring growing, you may be able to get ahead of weather with some strategic plantings. About two weeks before the last usual frost date in your area, dig the hill. Wait a week for the soil to resettle. Then plant a cluster of 4 seeds about 2 inches deep in the very center of that hill. One week later sow another clump at 12 o'clock near the hill's edge. In another week, plant another clump at 3 o'clock, and continue doing this until one of the sowings sprouts. Perhaps the first try won't come up, but at least one of these sowings will. A later-to-sprout position that experiences better weather in its first weeks may hugely outgrow seedlings that came up earlier. Thin gradually to the best one or two plants on the hill by the time the vines are running. If there are vine borers in your region growing two plants per position is a smart strategy.

The area around Elkton, Oregon where I lived has a longer and warmer summer growing season than most of Cascadia. There I attempted the first squash sowing about April 15. In the Willamette, May 1 works better. In the cooler parts of Cascadia, squash seed may not come up until June 1. In the humid eastern states it is a similar picture going from south to north. In the Middle South there is a long growing season and lots of heat; sow May through June. In the Deep South all forms of squash can be sown between March and the end of July, though later plantings are more subject to disease and insect pressure.

The amount of room to give each plant should be the same as the length of the variety's maximum vine growth. Most large fruited (*C.*

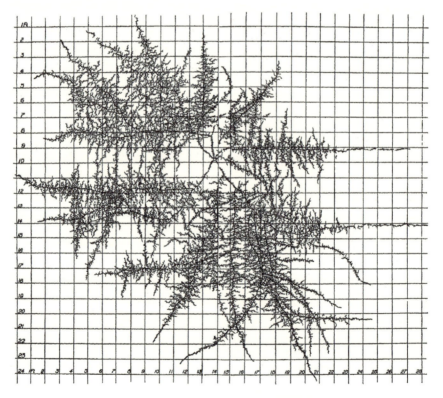

Winter squash (c. maxima) at maturity

maxima and some *C. moschata*) long-storing winter squash varieties can completely cover a 10-foot-diameter circle. Some spread even further. Butternuts make smaller vines. You can assume they'll spread over an 8 foot diameter. Small-fruited winter squash (*C. pepo*) like Delicata are even less sprawly—they grow to 5 to 6 feet across. Summer squash heirlooms like Early Yellow Crookneck can desiccate an 8- or 9-foot-diameter circle. Using abundant irrigation, fertilizer and drip-bucket fertigation. I once pushed a single *C. maxima* plant to cover a 20 foot diameter circle.

In Cascadia's extreme summer dryness an entirely unirrigated unfertigated vine yields at best 15 pounds of squash. However, support that vine with minimal fertigation every two to three weeks and you may harvest 60 pounds of squash from the same area. And that fertigated vine will cover more than twice as much ground. When fertigating squash vines, avoid compacting their root zone by standing in it or damaging vines where the bucket is set down; position drip buckets at the edge of the vine. The first fertigation may only need 2 gallons emptied into one position close to the hill. Then mid-July give two 3 gallon buckets per plant; about August 1, 8, and 15th, feed 10 gallons, dividing that up among three or four positions around the vine's perimeter. After that date, solar intensity and temperatures decline, growth rate slows, and water demand also decreases. On September 1 I'd add about 8 gallons and 4 more on September 15 if it hadn't yet rained significantly. Total water: 50 gallons. Total increase in yield: 45 pounds. I'd say that's a good return on water invested.

All vining winter squash varieties seem acceptably adapted to dry gardening. I wouldn't trust any of the newer compact bush winter squash varieties so popular on intensive raised beds.

The original warty Early Yellow Crookneck is an heirloom summer squash that grows enormous, high-yielding plants whose extent nearly rivals that of the largest winter squash varieties. It also forms a dense leaf cover that makes the fruit hard to find. Its fruit do not detach easily from the vine. The flavor of the original Yellow Crookneck is especially rich, probably due to its thick, oily skin; most gardeners who once grow the original Crookneck never again grow any other kind. Unfortunately, there are now "improved" EYCN varieties that do not sprawl as far, do not have such warty skin, have a brighter yellow skin color, are much easier to pick and do not have the same great flavor. The original Yellow Crookneck begins yielding several weeks later than the modern variant. However, as the summer goes on it will produce quite a bit more squash than modern types.

For those with space to spare I suggest only growing two early hybrid zucchini plants to fill the gap and a Yellow Crookneck. Soon your picking bucket will be filled with Crooknecks.

Tomato

There are two kinds of tomato growth habits: determinate and indeterminate. Determinate varieties make shorter vines that do not adapt well to staking or training. They should be allowed to sprawl. Most of them have been selected to keep (most of) their fruit from direct contact with the soil. If an indeterminate variety is grown that way it will form such a tangle of vines that it can be hard to even find ripe fruit, and far too many of them will be damaged, both from contact with the soil and by small soil-dwellers like sow bugs, earwigs and slugs. Dry gardeners should choose indeterminate varieties because their root systems extend at least as far as the vines are long—and the vines do not stop lengthening. As the vines grow I hold them up off the ground by tying them to wooden stakes. The stakes are positioned 18- to 24 inches apart along each vine in roughly concentric circles located wherever the vines wish to grow. As the vines lengthen I rigorously pinch off all weak side branches, which constitute two-thirds of them. By the time summer wanes there will be two and sometimes three rough rings of stakes around the center.

When dry gardening tomatoes the root system must be allowed to control all the space it can without any competition. Every summer I grow five indeterminate plants and give each one an uncontested 7 foot

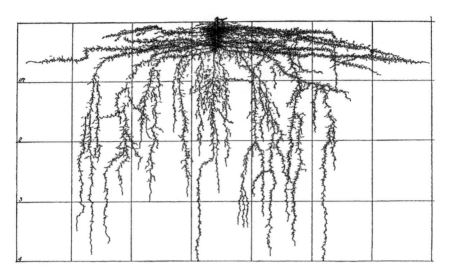

Tomato vine 8 weeks after transplanting

diameter circle for root zone. Because they are in rich soil and fertigated, by late summer it gets difficult to squirm between plants.

Spending time on staking and training has advantages. Very few fruit will show skin damage or evidence of being chewed by slugs or other primary decomposers. Because pinching off side branches effectively reduces the plants' fruit load, the tomatoes I harvest are a size larger than they might otherwise be. Maybe they taste a shade better too.

Set out transplants at the usual time. Spacing between plants depends on the vigor of the variety you're growing. In my experience, cherry tomato varieties cover the largest area and are most likely the very best dry gardening candidates—especially the true heirloom Red Cherry. Yellow Pear also grows an enormous plant. I am sure there are many other well-adapted recently developed cherry tomato varieties I can not access because of Australia Quarantine. I garden in a climate with cool summer nights that do not provoke the tomato to make maximum vegetative growth, so as I mentioned before, I assign a 7 foot diameter circle as uncontested root zone for each indeterminate plant. If I had to endure warm humid summer nights I could barely sleep in and had plenty of water I'd probably put each plant in the middle of a 10 foot diameter circle.

Periodic fertigation will greatly increase yield and the size of fruit even if it rains enough to keep the vines from suffering much moisture stress. Greenhouse tomato growers using hydroponics are constantly fertigating at a low intensity. They have learned to manipulate the plant's nutrition so that once it starts setting fruit they reduce nitrogen in order to slow vegetative growth and increase phosphorus and potassium in order to accentuate fruit set and development. If you fertigate regularly it would be possible to roughly accomplish the same thing outdoors. Before fruit starts setting generously fertigate the plant with NPK at the ratio of 1:1:1 plus all trace elements in balanced abundance. When fruit begins setting, change the NPK ratio to 0.5:2:2 and continue supplying trace elements. Using a product primarily intended for cannabis like DynaGro's "Grow" and "Bloom" accomplishes this, more or less.

Louisiana Extension says gardeners in the Deep South can set out tomato transplants in March/April and again July through mid-August for harvest late summer and into the cool season. My publisher says most tomato varieties in his garden do not withstand South Alabama's summer's heat and humidity. So I appreciate that two crops per year are a blessing. Given appropriate growing space there should be no need to be concerned about water in either season.

Which Vegetables To Focus On

As I wrote this chapter it became ever more clear to me which vegetables I'd concentrate on if I were trying to self-sufficiently produce as much of the family's diet as I could.

The Humid Eastern States
- Corn, field varieties for dry seed as well as sweet corn
- Potato
- Winter Squash, especially long keeping *C. maxima* varieties
- Yellow Crookneck and other summer squash

- Beans, snap, shelling and dry

- Garden pea (also garbanzo)

- Rutabaga

- Kale (both types)

- Lettuce (not during summer's heat)

- Savoy cabbage

- Carrot

- Beet

- Parsnip

- Parsley

- Tomato (indeterminate)

- Pepper (small fruited, both sweet and hot)

- Cucumber (especially Apple and Lemon)

Cascadia

Same as above, except that I'd eliminate field corn as a family staple and depend more heavily on potato and rutabaga.

Deep South

Same as above; add turnip, *C. moschata* pumpkins, lettuce in the cool season, collard, sweet potato, okra and cow peas. Dent corn usually does best as a grain, whereas flint corns generally do better in the north.

Chapter 6

My Own Garden Plan

All these photographs were taken at Elkton, Oregon late in July of 1992 when I was only 49 years old.

Dry gardened Gold Nugget tomatoes for lunch.

Romaine lettuce grows great during a hot summer—if it gets lots of water.

Winter Kale, tomatoes, and sunflowers. The row- and between-row spacing is about 4 feet.

Eggplants already! Notice the cloche support.
The plastic tunnel was removed over a month earlier.

Golden zucchini raised with abundant fertigation.

Carrots thinned to a foot apart. By October the carrots were 3 to 5 inches in diameter and a foot long.

Unirrigated kale thrives, but weaker-rooted Brussels sprouts were not fertigated and are stunted. The endives in the foreground grew crowded and lush by September, without rain or fertigation.

The sprinklers (left side of picture) spray only as far as the bush beans on the right side of the picture. Beyond the beans, summer squash (far right) is fertigated. I am watering a seedbed.

This chapter discusses my own dry garden plans and results from the early 1990s. My garden plan appears on the last page. Any garden plan is a product of compromises and preferences; my garden is not intended to become yours. But, all modesty aside, this plan results from the 20 continuous years of serious vegetable gardening I had done up to that time.

These days this old guy's dietary is "ketovegetablitarian". Not vegetarian, not lacto-ovo vegetarian. I eat some animal foods almost every day but do not focus on them. At least half of my calories come from nutrient-dense vegetables I grow myself. I eat small amounts of whole grain cereals. I avoid dairy (except butter, which I use liberally) because my body reacts badly to dairy protein. I avoid vegetable oils except olive and coconut. The purpose of my garden is to provide at least half the actual calories my wife and I eat year-round. Our bread is made from wheat, rye and spelt that we grind to flour shortly before baking. I try hard to put at least one very large bowl of leafy greens salad into my belly every day, winter and summer. I keep us in potatoes nine months a year and produce our full year's supply of onions. To break winter's dietary monotony I grow as wide an assortment of winter vegetables as possible. The appearance of my own vegetables puts most supermarket produce departments to shame when the summer veggies are "on". Their flavor and nutrient-density shames the industrial food system throughout the entire year.

My 1990s gardens may seem unusually large, but in accordance with Solomon's First Law of Abundance, there's a great deal of intentional waste so that moochers, poachers, guests, beloved friends, children, my own mistakes, poor yields, and failures of individual vegetables are inconsequential. Besides, gardening is fun.

The garden described in this chapter is laid out in 125-foot-long rows and one equally long four-foot-wide raised bed. (In the descriptions that follow and in the garden plan illustration, I've numbered the rows beginning on one side of the raised bed and moving outward, then going to the other side of the raised bed and moving outward again.) Each row grows only one or two types of vegetables. The central focus of my water-wise garden is on its irrigation system. Two lines of low output low-angle sprinklers straddle one semi-intensive raised bed running down the center of the garden. The sprinklers I used are a unique design. Each one emits only one gallon per minute and throws it at a very low angle so their maximum reach is about 18 feet; each sprinkler is about 12 feet from its neighbor in the row. On the illustration of the garden plan, the sprinklers are indicated by a circled x.

On the far left side of the garden plan illustration is a curved shaded area representing the uneven application of water put down by this sprinkler system. The 4-foot-wide raised bed gets lots of water, uniformly distributed. About half as much water lands only 6 feet from the edge of the raised bed as on the bed itself. Beyond that, the amount tapers off to insignificance. Crops are positioned according to their need for or ability to benefit from supplementation.

The Raised Bed

Crops demanding consistently moist soil are grown on the raised bed. These include a succession of lettuce plantings designed to fill the summer salad bowl, summer spinach, spring kohlrabi, my celery patch, scallions, Chinese cabbages, radishes, and various nursery beds that start overwintered crops for transplanting later. Perhaps the bed seems too large just for salads. But at that time one entire meal every day consisted largely of fresh, raw, high-protein green leaves; loose leaf or semi-heading lettuce is our main salad ingredient. These days my personal salad bowl still is larger than most families of six might consider adequate to serve all of them together.

Row 1: Succession Plantings
The row's center is about 3 feet from the edge of the raised bed. In March I sowed our very first salad greens down half this row—mostly assorted leaf lettuce plus some spinach—and somewhat later sowed six closely spaced early hybrid zucchini plants. The greens were all cut by mid-June; by mid-July my better-quality Yellow Crookneck squash came on (in another area), so I pulled the zucchini. Then I tilled that entire

row, refertilized, and sowed half to rutabagas. The nursery bed of leek seedlings had gotten large enough to transplant at this time, too. These go into a trench dug into the other half of the row. The leeks and rutabagas could have been reasonably productive located farther from the sprinklers, but no vegetables benefit more from abundant water or are more important to a self-sufficient kitchen. Rutabagas break the winter monotony of potatoes; leeks vitally improve winter salads, and leeky soups were a household staple from November through March.

Row 2: Semi-drought-tolerant Brassicas
Row 2 gets about half the irrigation of row 1 and about one-third as much as the raised bed, and so row 2 is 4 feet wide, including the path between that row and row 1. This gives the roots more room. One-third of the row grows savoy cabbage, the rest, Brussels sprouts. These brassicas are spaced 4 feet apart and by summer's end the lusty sprouts form a solid hedge 4 feet tall.

Row 3: Kale
Row 3 grows 125 feet of various kinds of kale sown in April. There's just enough overspray to keep the plants from getting gnarly. I prefer kale to not get very stunted, if only for aesthetics. On my soil, one vanity fertigation about mid-July kept this row looking impressive all summer. Other gardens with poorer soil might need more support. This much kale may seem an enormous oversupply during summer, but between salads and steaming kale with potatoes we manage to eat almost all the tender small leaves it grows during winter.

Row 4: Root Crops
Mostly carrots, a few beets. No irrigation, no fertigation, none needed. One hundred carrots weighing in at around 5 pounds each and 20-some beets of equal magnitude make our year's supply for salads, soups, and a little juicing.

Row 5: Dry-gardened Salads
This row holds a few crowns of French sorrel, a few feet of parsley. Over a dozen giant kohlrabi are spring sown, but over half the row grows

endive. I give this row absolutely no water. Again, when contemplating the amount of space it takes, keep in mind that this endive and grated kohlrabi must help fill our salad bowls from October through March.

Row 6: Peas, Overwintered Cauliflower, and all Solanaceae

Half the row grew early bush peas. Without overhead irrigation to bother them, unpicked pods form seeds that germinate with great vigor the next year. After the pea vines come out their half of the row is rotary tilled and fertilized again. Then it stayed bare through July while capillarity somewhat recharges the soil. About August 1, I watered a narrow row in the bed's center with hose and fan nozzle and sowed seeds for overwintered cauliflower. To keep the cauliflower from stunting I lightly hand sprinkled the row's center once a week through late September. Were water more restricted, I could have started cauliflower seedlings in a nursery bed and transplanted them here in October.

The other half of the row was home to the *Solanaceae*: tomatoes, peppers, and eggplant. I gave this row a little extra width because pea vines run, and I fertigated my *Solanaceae*, preferring sprawly tomato varieties that may cover an 8-foot-diameter circle. I left a couple feet of bare earth along the outside because the neighboring pasture grass deplete soil moisture along the edge of the garden.

Row 7: Water-demanding Brassicas

Moving away from irrigation on the other side of the raised bed, I grew a succession of hybrid broccoli varieties and late fall heading cauliflower. The broccoli was sown several times, 20 row-feet each sowing, done about April 15, June 1, and July 15. The late cauliflower went in about July 1. If necessary I could have used much of this row for a green manure or for quick crops that would be harvested before I wanted to sow broccoli or cauliflower, but I did not need more room. The first sowings of broccoli were pulled out early enough to permit succession sowings of arugula or other late salad greens.

Row 8: The Tellis

Here I erected a 125-foot-long, 6-foot-tall net trellis for gourmet delicacies like pole peas and pole beans. The bean vines blocked almost all water that would reach beyond it and so this row got more irrigation than it otherwise might. The peas were harvested early enough to permit a succession sowing of Purple Sprouting broccoli in mid-July. Purple Sprouting needs a bit of sprinkling to germinate in the heat of midsummer, but, being as vigorous as kale, once up, it grew adequately on the overspray from the raised bed. The beans would overwhelm our kitchen if all were sown at one time, so I planted them in two stages about three weeks apart. Still, a great many beans go unpicked. These were allowed to form seed, were harvested before they got entirely dry and then finished drying under cover away from the sprinklers. We got enough seed from this row for planting next year, plus all the dry beans we cared to eat during winter. Dry beans are hard to digest, and as we aged we ate fewer and fewer of them.

Row 9: Cucurbits

This row was so wide because it contained all the spreading cucurbits. The pole beans in row 8 tended to prevent overspray; this dryness was especially beneficial to humidity-sensitive melons, serendipitously reducing their susceptibility to powdery mildew diseases. All cucurbits were fertigated every three weeks. The melons were done by mid-September and the squash fell apart by the end of September. Once these were pulled, the area was tilled and fertilized, making space to transplant overwintered spring cabbages, other overwintered brassicas, and winter scallions in October. These transplants were dug from nurseries on the irrigated raised bed. I could also have set cold frames here and forced tender salad greens all winter.

Row 10: Unirrigated Potatoes

This single long row satisfied a potato-loving household all winter. The quality of these dry-gardened tubers was so high that my wife complained when she had to buy a few new potatoes from the supermarket after our supplies had become so sprouty and/or shriveled that they were no longer tasty.

Part II

Gardeners Textbook of Sprinkler Irrigation

Anyone who *can* irrigate their vegetable garden will choose to do so. They will irrigate before the top 6 inches gets dry. They will frequently water young seedlings until they put roots into soil that remains moist after two hot sunny days. They will wet down beds of germinating small seeds that can not be deeply buried. That's why I am certain some readers will start reading this book here and not at the beginning. For their benefit I will briefly repeat a few things I discussed in Part 1. If you already read the first half of this book I ask your patience.

Most gardeners irrigate without understanding the science behind it. Consequently, many spread too much water and thereby leach their soil, slow growth and reduce their yields. So listen up now! Here is irrigation science made as simple as I can make it.

Chapter 7

General Principles

Soil is a made up of tiny rock fragments (sand and silt) and clay— which is not a rock particle but actually forms in the soil out of chemicals released from decomposing rock fragments. One or a combination of all of these three components typically forms at least 92 percent of the solid soil mass. Soil also contains fully- and partly decomposed organic matter, water, air and living things ranging in size from bacteria up to worms and insects. In hot semi-arid and arid climates there is usually so little organic matter and so few living things that the mineral components consist of 98% to 99% of the total soil mass.

Smart gardeners go to great lengths to increase the amount of organic matter in soil. When there is more SOM most kinds of soil develop a stable structure that includes many small openings, gaps and spaces so it will hold as much air as possible. This is a very good thing because one of the most important plant nutrients in soil is oxygen. The details of how to increase the soil's organic matter and air supply belong in another book.

When it rains hard or when the irrigation system spreads water as fast or faster than the soil can assimilate it, the soil's air spaces temporarily fill up with water. Air is temporarily driven out. That's a good thing as long as the airless condition only lasts hours and not days because soil air contains less oxygen and a great deal more carbon dioxide than the atmosphere does. The CO_2 was exhaled by the soil biota and plant root cells. When the moisture inflow stops, excess water drains from the soil's pore spaces and fresh oxygen-rich air gets sucked back in. Then plants grow better for awhile. Only part of that improvement is caused by the plants having abundant moisture. Partly it is due to there being more soil oxygen because roots are made of living cells that need oxygen to remain alive and conduct their business. Business? The roots do much more than forage for water. Root cells also manufacture growth regulators and phytamin-like substances that rise into the entire aboveground plant.

After the rain ends or irrigation stops, not all the moisture they brought drains away; much of it adheres to soil particles in much the same way a balloon charged with static electricity sticks on a wall. To see *adhesion* at work, dip a small stone into a glass of water and then remove it. A few drops may fall off and then the wet rock stops dripping. If this rock were soil, then the amount of water adhering to its surface at the moment dripping ceases would be its *capacity* to hold water. At that moment the rock could be said to be at *full capacity*.

When soil moisture is at, or close to full capacity plants can extract water without much effort. As soil dries down the film of water adhering to its particles gets ever thinner, and the thinner it becomes the harder it sticks and the more energy plants must use to extract that moisture. This effort is called *moisture stress*. Moisture stress slows growth because it uses up plant energy that otherwise would have been used for growth. For that reason alone moisture stress is something the gardener wants to minimize. Moisture stress also makes lettuce taste bitter. If a moisture-stressed cauliflower manages to form a head at all, then it will be harsh-tasting and lack crispness. Moisture-stressed zucchini become dry, pithy and sometimes bitter; moisture-stressed cucumbers often are bitter. Moisture-stressed squash vines don't set as many fruit and they'll be smaller and less tasty. Radishes become dry, chewy and pungent to the point of bitterness. Celery hardly grows at all and becomes unpleasant to eat. I could extend this list but I think you got my message.

Most kinds of vegetables wilt the moment they fail to extract enough soil moisture to keep up with the amount being lost in strong sunshine. Wilting is a self protective action that reduces moisture loss because drooping leaves no longer catch as much sunlight. The plants may recover when the day cools down or the sun goes behind thick clouds; they may look entirely okay the next morning. Actually, temporary wilting is a huge stress that slows growth for days after. Some crops store water—lettuce, endive, parsley, onion, radish, rocket, Swiss chard, beet, parsnip, carrot and cilantro. By drawing down their moisture reserve these species may pass through a moisture deficit without wilting (and without growing much, if any)—but carrot and radish become bitter and tough; parsley stops making new leaves, etc. No wonder gardeners quickly learn to fear dry soil.

On the other hand, spreading more water than has been lost through evapotranspiration relocates plant nutrients out of the topsoil and into the subsoil, where they may be out of reach of the crop's root system. The solution to both too much and too little soil moisture is irrigating systematically, replacing lost soil moisture before there is much moisture stress but not adding any more water than necessary. However, installing a costly irrigation system is not essential in the home garden. You can correctly water vegetables with hose and nozzle or inexpensive lawn sprinklers if you know what you want to accomplish and discover with a simple test how fast and how uniformly the lawn and garden sprinklers you already own spread water.

Chapter 8

Systematic Watering

Wetting down the garden with hose and nozzle can be made to work quite well—if you enjoy the task—and if the garden is small. But if you don't relish the occasion then the garden usually gets watered too briefly. If the garden is watered every day, but too briefly, the gardener may not realize that starting a few inches below the surface the bed may have been sucked bone dry. The gardener does not realize this is happening because the vegetables themselves never wilt; they're sustained by surface moisture. The gardener also never realizes they grow poorly because most of their potential root zone is unusable.

To avoid this possibility, John Jeavons, the main popularizer of the intensive method, recommends repeatedly directing the nozzle's spray back and forth across several square yards of bed (area A) until the entire surface sparkles, becomes shiny wet. The shine persists until the excess moisture has flowed deeper. Spreading even more water before the surface pore spaces drain would only have it run off the bed. Rather than wasting water (and risk washing away topsoil), move on to the next few square yards (area B) and continue on B until that section gets sparkly and then return to A until the shine reappears on A, and then back to B until the shine reappears, repeating this pattern until the sparkling shine lasts long enough on both areas. The first shine that appears only lasts a second or so but as moisture flows deeper the shine lasts longer. How long you want the shine to persist depends on your soil type and on how much organic matter it holds. The duration of the shine varies from one second on sand to ten seconds or longer on clay. That's quite a difference!

Jeavons suggests learning how your soil accepts moisture by trying different shine durations combined with digging test holes to see how deeply the soil had become saturated. Without this check, gross overwatering or underwatering could result. The desired shiny time only has to be determined once. Clayey soil holds a great deal of moisture but assimilates it slowly. So clay can show a sparkly surface for many seconds whilst still being quite dry a few inches down. On the other hand, coarse sand takes in water so rapidly it can be difficult to get a shine to appear no matter how fast water

> **Leaching:** When water from rain or irrigation passes through soil it picks up plant nutrients already dissolved in soil moisture and transports them as deeply as the water penetrates, where they remain. These nutrients have been *leached*. If nutrients are deposited deeper than the crop's roots reach, they have been made unavailable and the crop's root zone has been impoverished.

is put down— but sand gets leached readily. Some fine sand soils repel water when they get dry on the surface and rain or irrigation water runs off without penetrating. It can take quite a few minutes of repeatedly moistening the surface and waiting for the shine to disappear before the soil will accept moisture more rapidly. Typical of garden writers, Jeavons did his learning on a clay soil and proceeded to expand what he observed from that limited experience to include every situation everywhere. This exemplifies a typical garden writer's pitfall. I mention another example of this earlier in this book when I discuss the use of permanent mulch. That method works fine in Connecticut where Ruth developed the method— a place where winter is truly wintery, where the soil freezes solid for several months. But where winter is mild, such as Cascadia, such as California, permanent mulch soon leads to breeding up a plague of primary decomposers like slugs and sow bugs and earwigs.

Hand watering with hose and nozzle can be made to work well if you have time to spare. I started gardening this way but I have always grown a large garden and in those days I had a small business to run. So I soon switched to sprinklers that achieved a similar result without consuming my time.

A wet to dry scale

	Soil moisture remaining
Permanent wilting point	20–33%
Temporary wilting point	50%
Minimum moisture for intensive vegetable beds	70%
Field capacity	100%

Water needed to bring 12 inches of soil from 70 percent of capacity to full capacity

Soil type	Irrigation in inches
Sand	1/4 inch to 1/2 inch
Medium (loam)	5/8 inch to 1 1/4 inches
Clay	1 1/4 inches to 1 3/4 inches

Average summertime soil moisture loss, assuming the soil is covered by a dense leaf canopy

Type of climate	Loss per day
Cool dry climate (sunny, under 75°)	3/16 inch
Cool dry climate (sunny, over 75°)	1/4 inch
Hot humid climate (85°, sunny)	3/8 inch
Hot dry climate (+100°, breezy)	1/2 inch

Daily moisture loss varies with the season and with the location. The amount lost depends upon light intensity, temperature, wind, humidity and the amount of leaf cover. The scientific name for this loss is *evapotranspiration,* meaning evaporation from the soil's surface plus transpiration from leaves. Much more moisture transpires from plant surfaces than evaporates from bare soil. Much more. As a crop grows its leaves prevent sunlight from evaporating soil moisture and instead, the plants transpire far more moisture than would ever evaporate. So the rate of moisture loss increases as the crop grows.

Sandy soil supporting advanced crops that have root systems more than 1 foot deep should be given ½ inch of water every two days in average sunny summer weather. Sandy soil growing young plants (with shallow root systems) will grow them the fastest if it is given ¼ inch every morning when the previous day had been warm and sunny. Clayey soils supporting advanced crops can accept 2 inches of water without having plant nutrients leached below the root zone. This much every five or six days in summer is enough, except when temperatures go over 90° for several days in a row.

Irrigating commercial vegetable crops became the usual thing in the early twentieth century. Moisture stress became conveniently preventable. In response, plant breeders redesigned vegetables to grow quicker above ground at the expense of root system development. This brings about larger harvests sooner—as long as the soil is kept moist enough. That's why modern vegetable varieties, be they hybrids or open-pollinated, only grow well when soil is comfortably moist. So be guided by this principle: Once the topsoil has dried to about 70 percent of its total capacity to hold moisture, it should be brought back to capacity again. The top foot of an average sand soil that is supporting a full leaf canopy can go from full moisture capacity to 70 percent of capacity in two average sunny summer days. A garden on coarse sand may benefit from being topped up after only one hot, low-humidity breezy day. A clayey soil similarly covered with a crop canopy loses the same amount of moisture each day as sand does but this garden

> Squeeze a small handful of soil into a ball you can hold in your fist, about the size of a golfball. If the ball feels wet, gooey, or sticks together solidly then the soil moisture is well over 70 percent, unless it is sand; sand won't form a ball no matter how hard you squeeze. If the ball sticks together firmly but breaks apart easily with a little thumb pressure then moisture content is around 70 percent. If a ball won't form at all then moisture is below 65 percent.

could be watered every five to seven days, and you'd spread much more water with each irrigation.

New seedlings feed in the top few inches. When moisture measured a few inches deep has dropped to 70 percent of full capacity, irrigate these by spreading half the amount of water recommended in the table titled "Water needed to bring a 12 inches of soil from 70 percent of capacity to full capacity". When the crop is two months old (after one month if the vegetable makes a tap root) the root system reaches down two feet, even deeper than that if subsoil conditions permit. Within two months the leaves of vegetable crops form a dense canopy that transpires a lot of moisture. At this stage, replace about 10 percent more than the "Summertime soil moisture loss" according to the table on the previous page.

Some parts of the garden benefit greatly from almost daily light irrigation in hot sunny weather. This includes beds germinating seeds, beds growing small seedlings and beds growing species with unusually high moisture requirements such as radishes and celery. These extra needs are best supplied with hose and nozzle so I always locate nursery beds growing leek seedlings and my celery and radish beds close to the garden hose tap.

A Cascadian soil supporting a full leaf canopy loses, on average, 1 inch of water through evapotranspiration each week during the month of June. During July and much of August water loss in Cascadia averages from 1¼ inches to 1½ inches per week. Moisture loss can be more than that when a hot dry wind blows from east of the Cascades. From late August to the first half of September moisture losses are ¾ inch to 1 inch per week. Average soil moisture losses differ in other parts of the country. On page 6 I have included a table showing average soil moisture losses in different parts of the USA.

I determine when and how much to irrigate without actually testing soil moisture. I assume moisture loss has been about what the table "Average summertime soil moisture loss" located a few pages back says it is, and periodically spread 10 percent more than that amount, less any rain received since the last irrigation. I also assume that on drizzly days there is no moisture lost and that on overcast days the soil loses about half as much moisture as when the sun shines.

I assume sandy soil supporting advanced crops should be given ½ inch of water every two days in average sunny summer weather. Sandy soil growing young plants (with shallow root systems) should be given ¼ inch every morning if the previous day had been warm and sunny. Clayey soils supporting advanced crops can accept 1½ inches of water without being leached. This much every five or six days in summer is enough, except when it goes over 95° F. Some parts of the garden require almost daily light irrigation in hot sunny weather. This includes beds germinating seeds or growing small seedlings and beds growing species with unusually high moisture requirements such as radishes and celery. These extra needs are best supplied with hose and nozzle.

Chapter 9

Designing Irrigation Systems

Lawn sprinklers spread water thick and fast because sod facilitates rapid moisture penetration. Quick seems convenient. But how uniformly do lawn sprinklers spread water? And how much time does your sprinkler have to run in order to spread a predetermined amount? Before irrigating a veggie garden with a lawn sprinkler please discover its application rate and uniformity of distribution. Position a few primitive irrigation rate measuring gauges—such as empty tin cans or straight-sided drinking glasses—in different parts of the sprinkler's pattern. Position one gauge a few feet from the sprinkler itself, put another 2–3 feet inside the limit of its throw and two more in the middle between these two. Then operate the sprinkler for exactly 30 minutes when the wind is not blowing. After 30 minutes measure the amounts of water in these gauges. Calculate the average and multiply by two in order to derive the sprinkler's average application rate per hour. Also make note of how much difference there is from gauge to gauge.

In my experience, lawn sprinklers covering a circular area spread more than 1 inch per hour and deposit more than twice as much water closer to the sprinkler than they do towards the outer limit of its throw radius. Oscillating sprinklers, the sort that cover a rectangle, spread moisture at least that fast and tend to deposit much more water towards the outer limit where the sprinkler reverses, than it does over the central part of its coverage. Spot sprinklers and perforated hoses that spray water three to six feet high and to both sides usually put out an even higher rate with even less uniformity of coverage.

Have you ever seen water flowing along the street because a suburbanite has been watering their front lawn? How much leaching do you suppose happens in a sandy food garden when a lawn sprinkler runs longer than necessary? The tin can test also reveals how uniformly the sprinkler spreads water. Are all the containers filled equally, or typically, are the ones closer to the sprinkler holding considerably more water than those on the fringe of its coverage? The user can compensate for unequal distribution by running the sprinkler for less time in each position and using overlapping positions.

Hardware shops, garden centers and big box home suppliers sell impact sprinklers resembling those used by farmers but these usually have large nozzles that make for high application rates better suited to lawns and rarely distribute uniformly. They also are consumer merchandise that usually lacks durability. I suggest you buy agricultural-quality crop sprinklers from an industrial-level supplier that can provide optional nozzle sizes and a range of sprinkler sizes. I'll soon explain why for vegetable crops you'll want small sprinklers and small nozzles—and I will make specific recommendations.

Before buying any sprinkler, discover what your water pressure is because this factor may determine which sort of sprinkler, if any, will work properly. Municipal water pressure usually exceeds 35 pounds per square inch (psi) at the street water main so that the homeowner's pressure regulator can hold their house pressure at a steady 35 psi. Homestead well pumps produce at least 35 psi but some of that pressure gets lost in the pipes. One way to gauge the actual pressure at the garden tap is to see how far the type of adjustable hose nozzle that can make a strong smooth jet can throw water. At 35 psi the stream should reach 40 to 50 feet in a windless moment.

Agricultural grade impact sprinkler nozzles suitable for vegetable gardening should emit between $1\frac{1}{4}$ to $2\frac{1}{2}$ gallons per minute. To calculate the number of sprinklers a tap can supply at one time, first count how many seconds it takes to fill a 5 gallon bucket through a full length of garden hose with an adjustable nozzle on it jetting out its strongest, longest-throwing stream. The hose nozzle should be used for this test because it mimics the back pressure from an array of sprinkler nozzles. Before building a system that uses the entire amount the tap puts out, first see if the flow rate lessens greatly when someone takes a shower or fills the kitchen sink while the bucket is filling. Maybe the irrigation system should be designed so as to not require the entire flow. Or perhaps it is possible to organize a larger water supply.

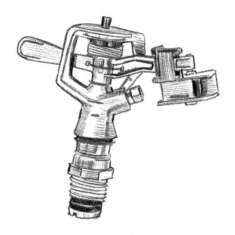

Rainbird 14VH. This is a high quality impact sprinkler

Agricultural sprinklers come in various designs and sizes; each sprinkler accepts a range of nozzle sizes. Farmer-sized sprinkler nozzles typically throw water 60+ feet. Some really big sprinklers with nozzle openings the size of a 10 gauge shotgun barrel can spray dairy manure slurry over a circle up to 100 yards in radius. The Rainbird 14VH is an impulse sprinkler designed for vegetable crops. It comes with nozzles between ¹/₁₆ inch and ⁷/₆₄ inch. A ¹/₁₆th inch nozzle can be expected to throw around 30 feet; a ⁷/₆₄ inch nozzle throws around 36 feet. Small-bore nozzles like this are better for growing vegetables because (1) they put out fine droplets that cause less soil compaction and reduce crust formation, and (2) a shorter throw radius helps keep the water off adjoining buildings and out of the neighbor's yard. Even if you have a

huge garden with heaps of ornamental plants or lawn around it that can
benefit from overspray, please avoid big sprinklers! Large droplets hit the
ground with enough impact to break up soil crumbs and float silt and
clay particles to the surface where they form a smooth crust, much the
same way screeding concrete lifts the fines to form a smooth skin on top.
Crust formation is the last thing you want; it blocks germinating seeds
and stops air exchange. Crusts rarely form on sand but sandy soil is easy
to overwater when the application rate is high. Use small nozzles!

Nozzle size

	1/16"		5/64"		3/32"		7/64"	
PSI	Rad.[1]	GPM	Rad.	GPM	Rad.	GPM	Rad.	GPM
20	29	0.59	30	0.79	33	1.14	34	1.55
25	29	0.56	31	0.88	33	1.27	35	1.73
30	29	0.62	31	0.97	34	1.39	35	1.90
35	30	0.66	32	1.05	34	1.50	36	2.05
40	30	0.72	32	1.12	35	1.61	37	2.19
45	31	0.75	33	1.19	35	1.71	37	2.32
50	31	0.80	34	1.25	36	1.80	38	2.45
55	32	0.84	34	1.31	36	1.89	38	2.57
60	32	0.88	34	1.37	37	1.97	38	2.68

Application rates from various nozzle sizes[2]

Nozzle bore	Pressure (psi)	Discharge (gal/min)	Throw	Spacing	Application (inches/hour)
1/16"	30	1.1	29'	20 x 20'	1/8
1/16"	60	1.6	32'	20 x 20'	3/16
1/8"	30	2.0	37'	20 x 20'	1/2
1/8"	60	2.9	40'	20 x 20'	9/16
3/16"	30	7.1	39'	23 x 26'	1.0
3/16"	60	10	46'	23 x 26'	1 1/4

[1] Throw radius in feet.

Application rates from various nozzle sizes[2]

Nozzle bore	Pressure (psi)	Discharge (gal/min)	Throw	Spacing	Application (inches/hour)
5/16"	30	20	59'	39 x 59'	11/16
5/16"	60	28	72'	39 x 59'	1.0

Application rates up to ½ inch per hour are ideal for most food gardens because the droplets are small and do not cause soil compaction or crust formation. The best way is with an array of uniformly spaced small nozzles all running at once covering as much of the garden as possible. Small bore sprinklers do have limitations. Strong sun and/or wind, especially with high temperatures can evaporate much of a thin water stream before it hits the ground. This means that small-bore nozzles are best operated early in the morning or at night.

A heavy clay soil could be watered from bedtime to breakfast without leaching the root zone if the application rate is ⅛ inch to 3/16 inch per hour. Watering at night is reputed to cause disease. Actually, watering all night prevents disease by continuously washing bacteria and fungus spores off the plants before they can germinate. This principle is well understood by nurseries; they root cuttings by frequently misting them. What can harm plants is to stop watering around nightfall. This makes plants damp all night—ideal conditions for the multiplication of disease organisms.

For sand or light loam soils, design a system that spreads ½ inch to ¾ inch per hour so that the previous two days' moisture loss can be restored in the morning before the sun gets strong and the wind comes up. For hydrophobic sands, an application rate of ¼ inch per hour should not run off the surface.

---

[2] I have included some large nozzles that make economic sense to farmers but not to gardeners.

Chapter 10

Uniformity of Application

Oscillating lawn sprinklers lay down a rectangular pattern that covers less uniformly than most circular-pattern sprinklers. I suspect the reason is that the device that moves the spray arm pauses at the turnaround points, putting too much water at the ends of its pattern and too little above the sprinkler itself. Got one of these sprinklers already? You can find out for yourself what it does with a few empty tin cans and 30 minutes.

An impact sprinkler has a single rotating nozzle. It is impossible to design one that spreads water uniformly over its entire throw radius because it must deposit nine times more water over the farther part of its throw than it does near the center. And every point in between must get a different amount.

The formula to calculate the area of a circle is: $A = 3.14 \times r^2$. Imagine that water is being spread by a sprinkler with a 25 foot throw radius. The five foot radius circle located at the center of the sprinkler pattern has an area of 78.5 square feet ($3.14 \times 5 \times 5$). The area of a five-foot-wide ring at the limit of coverage is 707 square feet, calculated this way: the area of

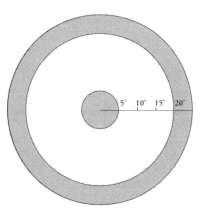

the full circle ($3.14 \times 25 \times 25$) minus the area in the inner 20 feet of the pattern ($3.14 \times 20 \times 20$). Thus the nozzle must deposit roughly nine times as much water over the outermost five feet of the pattern to end up with the same thickness of coverage as it spreads near the center.

The rocker arm that rotates the sprinkler head further prevents it from spreading water uniformly. It repeatedly interrupts the water jet, depositing considerably more water close to the sprinkler than further away.

Some consumer-market impact sprinklers come with a diffusing flap or an adjustable needle-tipped screw that breaks up the nozzle's stream and shortens the throw radius. But doing that makes it throw even less water to the fringes and even more water close in. The more the stream is interfered with the worse this effect becomes. Agricultural impulse sprinklers do not have diffusers; instead they come with precision nozzles that, if operated within the designed pressure range, put down only twice as much water near the center of their coverage as on the outer half.

Farmers compensate for unequal distribution by positioning sprinklers in an overlapping pattern—sometimes in a square pattern, sometimes hexagonal. The hex arrangement distributes water slightly more uniformly but the square pattern works better when the garden is close to buildings or where the sprinklers could throw water on the neighbor's property. This method works because the heavily watered area near one

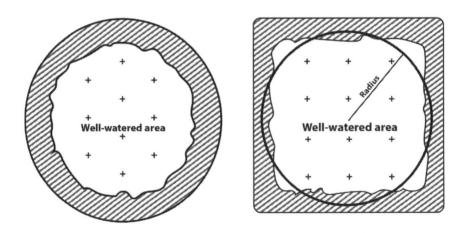

The gray shaded area gets some moisture, but not nearly as much as where many overlaps occur.

sprinkler is also covered by the lightly watered area of the next sprinkler in line.

High-angle sprinklers aim the stream farther above ground so they throw a greater distance. Rainbird high-angle sprinklers for example point the stream 23° above horizontal. High-angle sprinklers should be positioned no farther apart than 65 percent (⅔) of their throw radius. This results in the most uniform possible coverage and allows for (light) wind blowing the spray. Low-angle sprinklers throw at 10° above horizontal. The stream does not reach as far but is less affected by wind. Low-angle sprinklers can be separated by 75 percent of their throw radius. High-angle sprinklers with ¹/16th inch nozzles provide a very low rate of application that allows a clay soil to be irrigated all night, a time when there is much less wind.

Any multiple sprinkler pattern still leaves a dryish fringe where fewer over-laps occur. On the farm, throwing some water beyond the crop usually is of little consequence; in the backyard it may be essential to keep all spray within your own yard and/or off your own windows.

I like the useful and easily obtained sprinklers made by Rainbird (rain-bird.com). They manufacture two general designs. One is the 14VH impulse sprinkler I have already men-

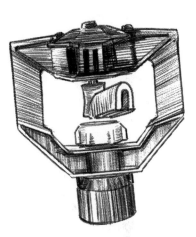

Rainbird LFX300

tioned. Two other designs are called "Low Flow". These are made of durable plastics. They accept a wide range of nozzle sizes that can be changed easily by the user. With the LFX design the nozzle jets the stream straight up; it then enters a spinner that redirects the stream close to horizontal and also spins at the same time. A range of nozzles provide flow rates from 0.3 to 0.7 gallons per minute; one spinner (diffuser) throws at 9° above horizontal, another throws at 15°.

Rainbird LF 1200

Rainbird also makes the LF design. This one is more costly and provides better coverage uniformity. The internal mechanism works like an impulse sprinkler.

The smallest model, the LF1200, has interchangeable nozzles ranging from 1 to 2.5 gallons per minute and deflectors adjusting throw angle between 6° up to 21°. A video on their website shows this model operating.

Tall growing crops like climbing beans, asparagus or sunflowers can be positioned to block water that would strike a building or go into a neighbor's yard. Some gardeners avoid spreading water outside the garden by using part-circle impact sprinklers. Keep in mind that cutting an impact sprinkler's arc in half doubles its rate of application; cutting it to a quarter-circle quadruples the rate of application.

Impact sprinklers spread water even less uniformly when they are running in part-circle mode because the rocker-arm's more rapid action when reversing drops a great deal of water close to the sprinkler. Part circle sprinklers come with other downsides. Actuating the reversing mechanism requires considerable force; this means a larger-bore nozzle putting out a powerful stream with high application rates and bigger droplets.

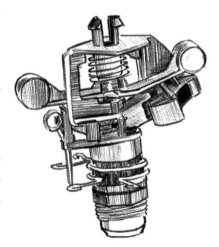

Rainbird 2045-PJ

Count on Rainbird to make good gear. Their 2045-PJ is a full- or part-circle impulse sprinkler that comes with a range of nozzles drawing from 1.5 gallons per minute up to 8.4 gpm, and corresponding throw radii from 23 feet to 45 feet.

Another way to reduce the arc of a full-circle sprinkler is to attach a shield to a garden stake pounded in directly behind it. The shield can be made from a cut-out tin can or half of a small plastic bucket. This method does not increase the application rate nor lessen uniformity of distribution—except for all the water dumped immediately behind the sprinkler.

In my opinion, when the gardener starts thinking they needs a part-circle impulse sprinkler, more likely what they really need is a turbine powered design. Although Rainbird makes this sort for landscapers, I think that Nelson (nelsonirrigation.com) makes better gear of this type. Turbine powered sprinklers use a slowly rotating circular nozzle head powered by an internal propeller; each nozzle head has numerous subnozzles, One subnozzle throws an undiffused stream that covers the outermost part; other jets spread water closer in. Some diffuse their stream; others make a rather clean stream. The end result is uniform coverage.

I have used Nelson's "MP Rotator" and so I can commend it for being reliable and not particularly costly. It allows handy adjustment of the stream angle (radius of throw). One sprinkler can be adjusted to cover less than a full circle at the same rate of application. Nelson's MP Rotator sprinklers come in several sizes providing different rates of application.

Nelson Rotator

Nelson makes another line of sprinklers they call simply "Rotator". These work much like the low flow Rainbirds I just described.

Agricultural sprinklers are designed to operate within a specific range of water pressures. If water pressure is too low for the design, then the nozzle emits a very clean, smooth stream that puts far too much water on the outer few feet of the pattern. Too little water

goes into the mid-zone. The resulting distribution pattern is termed "doughnutting" because it resembles the sweet breakfast pastry North Americans like.

Consider the opposite. Operated at excess pressure, the water jet becomes too turbulent, which actually shortens the throw while greatly increasing the amount laid down close to the sprinkler, making the fringes too dry. Overpressure also shortens the sprinkler's operating life.

Inexpensive sprinkler supports capable of supplying a small nozzle can be made by gluing a plastic micro sprinkler spike stand into the end of a ¾ inch white plastic water pipe that has its bottom end cut off at a sharp angle so that it can more easily be pushed into soft soil. The white pipe carries no water.

I urge all gardeners to visit an agricultural irrigation supplier, spend some time there; handle the bits sold there; study the charts and tables on a sprinkler manufacturers website and see if all the pieces don't fall into place.

If having a permanent sprinkler system that waters the entire garden or large parts of the garden at one time is beyond your budget then uniform irrigation can still be accomplished with only one impulse sprinkler mounted on a three-foot-tall homemade stand that is supplied by a hose. It is run for the same amount of time in each position you would have put a dozen sprinklers had you been able to afford them or had enough water flow to operate them all at once. I watered Territorial Seed Company's trials ground this way for the first few years. I made the stand with one sack of redimix concrete, a 5 gallon white plastic bucket, $1/2$ inch galvanized iron pipe and a few fittings.

Chapter 11

Drip

From 1980 through 1986 I had to supply my household, irrigate a three-quarter-acre variety trials ground and a 5,000 square foot kitchen garden—all from a three gallon per minute well. Three gallons per minute can't supply even one sprinkler nozzle large enough to be used when the sun is shining and the wind is blowing. So as soon as the seed business could afford the cost I switched the trials ground over to drip tapes—lightweight flexible plastic pipes with a pinhole emitter every foot.

Then I came to know way too much about drip—and here are the downsides. Drip tapes are expensive even when purchased in 2,500 foot long spools but it really didn't matter much how expensive my trials ground was to operate—it was producing priceless information that made all my customers gardens grow better. In addition to expense, there were other down points. No matter how careful I was, a few times each year I'd cut a dripline when hoeing weeds. Once cut these tubes couldn't be mended without leaving quite a big section of the line without a water supply. For one cut I'd have to replace 100 feet of dripline with a crop growing over it, not so easy to accomplish sometimes. All the drip emitter holes must face away from the soil or they may become plugged by soil particles being drawn back into the emitter hole when the water is turned off. But drip tape twists and rotates as it expands and contracts with changes in temperature. To minimize rotation and keep the water on young seedlings the drip tapes must be pinned firmly to the earth every few feet of run; and even so, many of the fine vertical streams of water still point to the side as much as straight up. Although I filtered the water supply, every tiny emitter hole on every line still had to be checked each and every time a line was turned on because an emitter hole could have become blocked by soil that was sucked back into the opening when the water was last shut off.

Drip tapes, individual drip emitters, and soaker hoses do not suit sand because in sand the water pretty much goes straight down without wicking out horizontally, leaving areas of totally dry soil. The only way around this on sand is to lay parallel drip lines 1 foot apart with emitters 1 foot apart; doing that gets outrageously expensive and greatly interferes with hoeing. If the soil is a medium loam or clay then water wicks out horizontally as much as 2 feet to either side of the drip line. Highly UV-resistant (long lasting) supply lines with emitters designed so they can not get plugged might be useful for permanent plantings such as berries, but with a large enough water supply, I'd always choose sprinklers.

Microirrigation is a hybrid between drip and crop sprinklers, using low-pressure black plastic pipes, quick-connect fittings and cheap plastic spike stands holding miniature sprinkler heads that throw six feet at best, with emission rates down to a few quarts per hour from each nozzle. Microirrigation provides an inexpensive way to establish orchards, vineyards, and for watering narrow ornamental beds around houses. Microirrigation parts are sold by the tens or dozens bubble packed in garden centers, but you'll find a broader range, probably at lower prices, at irrigation suppliers. If you're considering microirrigation don't assume the rate of application is low. Each micro sprinkler emits very little water, but the nozzle doesn't throw far. And use a very effective line filter. The nozzles are minuscule.

Chapter 12

Plant Spacing Possibilities in Relation to Soil Depth and Irrigation

If you will spend awhile contemplating this chart and the footnotes accompanying it, all the parts of this book may link together for you in a more useful way. This chart was not developed from years of double blind scientific trials. It mostly is my own speculation, imagination, good guesses and intuition. It may not apply quite so well outside of Western Oregon and Washington states and the Lower Mainland and Islands of British Columbia. That does not mean it will be of no use to imaginative people living elsewhere.

Plant spacing possibilities (in inches)

Potential rooting depth[1] Frequency of irrigation[2]	1 foot 2–3 days	2 foot 5–7 days	3–4 feet 2 weeks	4–6 feet 3–4 weeks	6–10 feet 8–12 weeks
Asparagus	not likely	18 x 60	18 x 60	18 x 60	18 x 60
Broad bean, autumn sown[3]	8 x 18	8 x 24	12 x 30	12 x 30	12 x 30
Bean, bush, snap	6 x 18	8 x 24	12 x 36	16 x 48	18 x 48
Bean, pole, snap	8 x 48	10 x 48	12 x 48	16 x 60	24 x 72[4]
Bean, runner	36 x 36	36 x 36	36 x 48	48 x 60	48 x 60[4]
Beet	3 x 18	4 x 18	4 x 24	6 x 30	12 x 48
Broccoli (Italian)	18 x 24	24 x 24	30 x 30	42 x 42	unlikely
Brussels sprouts	24 x 24	24 x 30	30 x 30	36 x 36	unlikely
Cabbage, Chinese[5]	18 x 18	18 x 24	unlikely	unlikely	not possible
Cabbage, early[3 5]	18 x 18	18 x 24	24 x 30	unlikely	not possible
Cabbage, midseason	24 x 24	24 x 24	30 x 30	36 x 36	don't know
Cabbage, autumn/winter	24 x 24	24 x 30	30 x 36	36 x 48	36 x 48
Carrots[6]	2 x 18	2 x 18	2–3 x 24	4 x 36	12 x 48
Cauliflower, spring sown[5]	24 x 24	24 x 24	30 x 36	Not possible	

Plant spacing possibilities (in inches)

Potential rooting depth[1] / Frequency of irrigation[2]	1 foot / 2–3 days	2 foot / 5–7 days	3–4 feet / 2 weeks	4–6 feet / 3–4 weeks	6–10 feet / 8–12 weeks
Cauliflower, autumn	24 x 24	24 x 30	30 x 36	36 x 42	don't know
Cauliflower, overwintered	24 x 24	24 x 30	24 x 30	matures on rainfall	not possible
Celery[5]	24 x 24	24 x 30	30 x 36	unlikely	not possible
Celeriac[5]	24 x 24	24 x 30	30 x 36	unlikely	not possible
Corn	8 x 36	8 x 36	10 x 42	24 x 48	48 x 48
Cucumber[6]	36 x 48	42 x 48	48 x 48	60 x 60	60 x 60[4]
Eggplant	24 x 24	24 x 30	30 x 36	36 x 48	36 x 48[4]
Endive (chicories)[3]	12 x 18	12 x 18	18 x 24	24 x 36	24 x 48
Garlic	6 x 12	6 x 15	6 x 18	matures on rainfall	
Kale	18 x 24	24 x 24	30 x 30	36 x 48	60 x 60
Kohlrabi, autumn	4 x 18	6 x 18	6 x 18	8 x 24	does not apply
Leeks	4 x 24	4 x 24	6 x 30	8–12 x 30	uncertain
Lettuce, iceberg	16 x 18	16 x 20	20 x 24	unlikely	not possible
Lettuce, leaf	8 x 18	12 x 18	14 x 18	becomes bitter	

Plant spacing possibilities (in inches)

Potential rooting depth[1] Frequency of irrigation[2]	1 foot 2–3 days	2 foot 5–7 days	3–4 feet 2 weeks	4–6 feet 3–4 weeks	6–10 feet 8–12 weeks
Melons[6]	48 x 48	60 x 60	72 x 72	72 x 96	72 x 96[4]
Mustard, autumn[10][3]	6 x 18	6 x 18	8 x 18	12 x 24	does not apply
Onions, bulbing	4 x 18	4 x 18	6 x 18	8 x 24	unlikely
Parsley[7]	6 x 18	6 x 18	6 x 24	8 x 30	12 x 36
Parsley root	6 x 18	6 x 18	6 x 24	8 x 30	unlikely
Parsnip	impossible	4 x 18	6 x 18	6 x 24	uncertain
Peas, climbing, in spring	4 x 36	4 x 36	4 x 36	4 x 36	uncertain
Peas, bush, in spring	1 x 18	1 x 18	2 x 24	2 x 24	uncertain
Peppers	18 x 24	24 x 24	30 x 36	36 x 48	36 x 48[4]
Potatoes	12 x 42	12 x 42	15 x 48	18 x 48	20 x 60
Radish, small[3]	2 x 12	2 x 12	requires abundant moisture		
Radish, winter[3]	4–6 x 18	4–6 x 24	8 x 24	uncertain	
Rutabaga	6 x 24	8 x 24	12 x 30	18 x 36	18 x 48
Salsify	impossible	5 x 18	6 x 18	10 x 24	12 x 24

Plant spacing possibilities (in inches)

| Potential rooting depth [1] | 1 foot | 2 foot | 3–4 feet | 4–6 feet | 6–10 feet |
Frequency of irrigation [2]	2–3 days	5–7 days	2 weeks	3–4 weeks	8–12 weeks
Scallions	0.5 x 18	0.75 x 18	1 x 18	1.5 x 18	unlikely
Spinach, summer	2 x 18	4 x 18	6 x 18	8 x 24	unlikely
Squash, bush (zucchini)	30 x 36 [8]	36 x 42	42 x 48	48 x 72	60 x 72 [4]
Strawberry	8 x 18	12 x 18	16 x 24	not possible	
Sunflower, giant	12 x 24 [8]	12 x 36 [8]	24 x 42	24 x 60	30 x 72
Swiss chard	12 x 18	12 x 24	16 x 30	18 x 36	24 x 48
Tomato, determinate, compact	24 x 36	36 x 36	42 x 42	48 x 48	unlikely
Tomato, indeterminate	24 x 36	36 x 48	48 x 48	60 x 60	72 x 72 [4]
Turnip, autumn [3]	3 x 18	5 x 18	8 x 18	unlikely	
Watermelon	72 x 72	84 x 84	96 x 96	96 x 120	96 x 120 [4]
Winter squash [9]	60 x 60 [8]	84 x 84	96 x 96	120 x 120	120 x 144 [4]

Notes

1. Impenetrable clay subsoil may still be useful in that it holds a lot of moisture that slowly rises up into the topsoil. Gravel below the topsoil holds no moisture, interrupts capillary flow and won't allow root penetration.
2. If it doesn't rain. As a rough rule of thumb, moisture loss in the Willamette Valley is approximately ¼ inch per day in hot weather assuming the crop has developed a full leaf canopy.
3. Early spring and late summer-sown crops may be crowded a bit because they are not likely to experience long spells of hot rainless weather, nor will they have a long time to grow fast and get big.
4. Fertigate every three weeks.
5. Not capable of dealing with dryish soil.
6. Thin progressively.
7. Spacing depends much on variety and location; expanse of roots = expanse of vines.
8. Will resume leaf growth shortly after long drought ends.
9. On shallow soil grow compact "mini" varieties.
10. Spacing depends on variety's vigor; could range from 8–12 feet under irrigated situations.

More Reading

Carter, Vernon Gill, and Tom Dale. *Topsoil and Civilization*. Norman, Okla.: University of Oklahoma Press, 1974. The history of civilization's destruction of one ecosystem after another by plowing and deforestation, and its grave implications for our country's long-term survival.

Cleveland, David A., and Daniela Soleri. *Food From Dryland Gardens: An Ecological, Nutritional and Social Approach to Small-Scale Household Food Production*. Tucson: Center for People, Food and Environment, 1991. World-conscious survey of low-tech food production in semiarid regions.

Faulkner, Edward H. *Plowman's Folly*. Norman, Okla.: University of Oklahoma Press, 1943. This book created quite a controversy in the 1940s. Faulkner stresses the vital importance of capillarity. He explains how conventional plowing stops this moisture flow.

Foth, Henry D. *Fundamentals of Soil Science*. Eighth Edition. New York: John Wylie &. Sons, 1990. A thorough yet readable basic soil science text at a level comfortable for university non-science majors.

Hamaker, John D. *The Survival of Civilization*. Annotated by Donald A. Weaver. Michigan/California: Hamaker-Weaver Publishers, 1982. Hamaker contradicts our current preoccupation with global warming and makes a believable case that a new epoch of planetary glaciation is coming, caused by an increase in greenhouse gases. The book is also a guide to soil enrichment with rock powders.

Kourik, Robert. *Drip Irrigation for Every Landscape and All Climates*. Santa Rosa, California: Metamorphic Press, 1992. A thorough manual of deep irrigation, full of technical and design information.

Nabhan, Gary. *The Desert Smells Like Rain: A Naturalist in Papago Indian Country*. San Francisco: North Point Press, 1987. Describes regionally useful Native American dry-gardening techniques.

Russell, Sir E. John. *Soil Conditions and Plant Growth*. Eighth Edition. New York: Longmans, Green & Co., 1950. Probably the finest, most human soil science text ever written. Russell avoids unnecessary mathematics and obscure terminology. I do not recommend the recent in-print edition, revised and enlarged by a committee.

Smith, J. Russell. *Tree Crops: A Permanent Agriculture*. New York: Harcourt, Brace and Company, 1929. Smith's visionary solution to upland erosion is growing unirrigated tree crops that produce cereal-like foods and nuts. Should sit on the "family bible shelf" of every permaculturalist.

Solomon, Steve J. *Growing Vegetables West of the Cascades*. Seattle: Sasquatch Books. The complete regional gardening textbook.

Organic Gardener's Composting. Portland, Ore.: George van Patten Publishing, 1992. Especially useful for its unique discussion of the overuse of compost and a non-ideological approach to raising the most nutritious food possible. This book is currently out of print. Good Books is planning to bring it back into print soon.

Stout, Ruth. *Gardening Without Work for the Aging, the Busy and the Indolent*. Old Greenwich, Conn.: Devin-Adair, 1961. Stout presents the original thesis of permanent mulching.

Turner, Frank Newman. *Fertility, Pastures and Cover Crops Based on Nature's Own Balanced Organic Pasture Feeds*. San Diego: Rateaver, 1975. Reprinted from the 1955 Faber and Faber edition. Organic farming using long rotations, including deeply rooted green manures developed to a high art. Turner maintained a productive organic dairy farm using subsoiling and long rotations involving tilled crops and semi-permanent grass/herb mixtures.

Van der Leeden, Frits, Fred L. Troise, and David K. Todd. *The Water Encyclopedia, Second Edition*. Chelsea, Mich.: Lewis Publishers, 1990. Reference data concerning every possible aspect of water. Available online from several websites; search Google to find it.

Weaver, John E., and William E. Bruner. *Root Development of Vegetable Crops*. New York: McGraw-Hill, 1927. Contains very interesting drawings showing the amazing depth and extent that vegetable roots are capable of in favorable soil.

Widtsoe, John A. *Dry Farming: A System of Agriculture for Countries Under Low Rainfall*. New York: The Macmillan Company, 1920. The best single review ever made of the possibilities of dry farming and dry gardening, sagely discussing the scientific basis behind the techniques. The quality of Widtsoe's understanding proves that newer is not necessarily better.

About the Author

Steve Solomon is the author of six food gardening books; the first was published in 1980. All but one of them are still in print. His first book, *Growing Vegetables West of the Cascades,* has been through six heavily revised editions and still serves as the food gardener's bible in Western Oregon and Washington states. He founded soilandhealth.org, a major online library supplying free book downloads focused on the foundations of the organic farming and gardening movement and a form of natural healing called Natural Hygiene in North America or Nature Cure in the UK. Steve must take his own medicine because at age 79 he and his wife Anne (age 81) reside on a beef, milk and vegetable producing 11-acre homestead on the island of Tasmania. Steve also moderates two internet e-mail discussion groups. He can be found every Sunday morning and most every Wednesday night at the Western Tiers Mennonite Church. You may contact him at stsolomo@internode.on.net.

Made in United States
Orlando, FL
09 May 2022

17676203R00114